Geoengineering, the Anthropocene and the End of Nature

Jeremy Baskin

Geoengineering, the Anthropocene and the End of Nature

Jeremy Baskin
Melbourne School of Government
University of Melbourne
Melbourne, Australia

ISBN 978-3-030-17358-6   ISBN 978-3-030-17359-3 (eBook)
https://doi.org/10.1007/978-3-030-17359-3

© The Editor(s) (if applicable) and The Author(s), under exclusive licence to Springer Nature Switzerland AG 2019
This work is subject to copyright. All rights are solely and exclusively licensed by the Publisher, whether the whole or part of the material is concerned, specifically the rights of translation, reprinting, reuse of illustrations, recitation, broadcasting, reproduction on microfilms or in any other physical way, and transmission or information storage and retrieval, electronic adaptation, computer software, or by similar or dissimilar methodology now known or hereafter developed.
The use of general descriptive names, registered names, trademarks, service marks, etc. in this publication does not imply, even in the absence of a specific statement, that such names are exempt from the relevant protective laws and regulations and therefore free for general use.
The publisher, the authors and the editors are safe to assume that the advice and information in this book are believed to be true and accurate at the date of publication. Neither the publisher nor the authors or the editors give a warranty, express or implied, with respect to the material contained herein or for any errors or omissions that may have been made. The publisher remains neutral with regard to jurisdictional claims in published maps and institutional affiliations.

Cover illustration: © Michael Portley

This Palgrave Macmillan imprint is published by the registered company Springer Nature Switzerland AG.
The registered company address is: Gewerbestrasse 11, 6330 Cham, Switzerland

To Sundhya for her love, friendship and critical advice through six long PhD years. And to Ananya whose presence reminds me daily that it really does matter what world we pass on to our children.

Preface

Solar geoengineering (SGE) is an imagined technology which proposes spraying sulphuric acid aerosols into the stratosphere in order to reflect incoming sunlight back into space and thereby reduce or slow global warming. It does not yet exist as a deployed technology. But it is being actively researched and advocated for, especially in the USA.

When I first came across solar geoengineering in the mid-2000s, I was inclined to regard it as both 'mad' and 'bad'. It sounded like a crazily impractical idea seemingly drawn from the realm of science fiction, and embracing the least attractive features of modern techno-science. I soon discovered the idea was being taken extremely seriously, in part because the observed cooling effects following major volcanic eruptions were seen as proof that it could 'work'. Volcanos. Science fiction.

The devastating volcanic eruption at Tambora in 1815 in what is now Indonesia caused global temperatures to plummet and led, directly and indirectly, to a death toll likely in the millions (Wood 2014). In much of the Northern hemisphere, the following year was known as 'The Year without a Summer'. The teenage Mary Shelley (daughter of the political philosopher and proto-anarchist William Godwin and the philosopher and early feminist Mary Wollstonecraft) spent that June in Switzerland with the young writers Percy Shelley, Lord Byron and John Polidori. Kept indoors by the unseasonably cold and inclement weather, they embarked on a competition as to who could write the best horror story.

Mary Shelley's contribution was *Frankenstein: The Modern Prometheus*, arguably the first work of science fiction. It tells of how Dr. Victor Frankenstein uses his knowledge of science and an obsession with simulating natural wonders, to impart life to non-living matter. In the laboratory, he aims for a beautiful creation but instead gives life to what he regards as a hideous creature, Frankenstein's monster. Repulsed by his creation Victor leaves and the abandoned monster flees, only to kill in rage as he is rejected by society.

One familiar reading of *Frankenstein* sees it as a cautionary tale about hubris and high technology. Dr. Frankenstein tried to act as God. He disrupted the natural order and created a monster leading to the ultimate downfall of both scientist and monster. But Bruno Latour argues, in 'Love Your Monsters', that Frankenstein's sin was rather '… that he abandoned the creature to itself' and failed to take responsibility for his own creation. 'Our sin is not that we created technologies but that we failed to love and care for them' (2012: n.p.).

Perhaps Latour is correct when it comes to taking responsibility for technologies that 'we' have created, that are already 'out there'. It does not, of course, follow that it is permissible to create any new technology so long as we promise to 'love and care' for it through good times and bad. Unreflective technophilia is not a virtue, even if technophobia is a vice. At the time of writing, solar geoengineering is still a nascent, imagined, technology for addressing climate change. It has not yet come into being as a deployed technology. Is solar geoengineering a 'monster'? Is it one 'we' should create? Is it one we can ever love?

These are complex questions, resistant to simple answers, and with contested values at their heart. In digging deeper, it has become apparent to me that geoengineering throws up more questions than answers, not just about climate and technology, but also about hubris and power. It troubles our 'common sense' about environmentalism. It also raises profound, even existential, questions about both the human and the more-than-human, and the relationship between them.

As I was finalising the thesis on which this book is based, two leading proponents of solar geoengineering, both based at Harvard, published an opinion piece with the apposite title 'Toward a More Reflective Planet'. It was an argument for literally making the planet reflect more sunlight using SGE: "we need to turn down the heat—and fast" (Keith and Wagner 2016: 2). This book also argues for a 'more reflective planet' but in a diametrically opposed sense. For me SGE provides a lens through which to reflect on our planet and our Earthly condition and our ethical obligation to do so. My objective is to critically interpret SGE, exploring what is at stake in its emergence, what is constraining its normalisation, and reflecting upon and unmasking, where necessary, the masking, reflective technology which is SGE.

The affect and the visceral and instinctive responses which so often accompany any reflection on geoengineering means that I am regularly pressed to give my view on whether solar geoengineering *should* be pursued and whether it *will* be pursued. My answer tends to be 'No' and 'Probably' respectively. This reflects the view I have come to, that solar geoengineering is 'bad' and 'sad', but imprudent rather than 'mad'. But these are *not* the questions posed in this book. My quest is how best to understand geoengineering, raising questions like 'what work does the idea do and what is at stake in its adoption?', 'for whose benefit is geoengineering imagined?' and 'what is constraining its emergence?', rather than 'is it bad?'. These are the types of questions that need to be answered prior to making any decision about proceeding with solar geoengineering.

Melbourne, Australia   Jeremy Baskin

References

Keith, D., & Wagner, G. (2016, June 16). Toward a more reflective planet. *Project Syndicate*. Retrieved January 20, 2019, from https://www.project-syndicate.org/commentary/albedo-modification-research-and-development-by-david-keith-and-gernot-wagner-2016-06

Latour, B. (2012). Love your monsters. *Breakthrough Journal*, 2. Retrieved January 20, 2019, from http://thebreakthrough.org/index.php/journal/past-issues/issue-2/love-your-monsters

Acknowledgments

This book began life as a PhD thesis although the ideas have developed and morphed substantially in the process of writing this monograph. I acknowledge, with heartfelt gratitude, the feedback and advice from my two supervisors at the University of Melbourne, Professor Robyn Eckersley, my primary supervisor, and Professor Ghassan Hage. I also acknowledge the helpful comments from the two anonymous examiners of that thesis, as well as from Sundhya Pahuja. I had the good fortune to spend the last quarter of 2016 with Professor Sheila Jasanoff at Harvard University as a Fellow of the Program on Science, Technology & Society. This was not only intellectually challenging, but it also gave me a front-row view of the active solar geoengineering programme being run out of Harvard. I have been able to present some of the content of this book at various conferences including the 4S and EASST conferences in 2018, the Environmental Justice Conference in Sydney in 2017, the SDN conference and the STS Roundtable at Harvard in 2016, and the first Climate Engineering Conference in Berlin in 2014. Questions, criticisms and conversations during these events have, hopefully, sharpened my arguments. Of course, in relation to all the above, the usual disclaimers apply and I take full responsibility for any shortcomings in the current text. On the production side I'd like to thank Sylvia Seldon for helping with various tasks including proof-reading and reference checking; as well as Rachael Ballard and Joanna O'Neill from Palgrave for commissioning the book and helping see it through to production. I wish to acknowledge the following for their kind permission to reproduce images: the family of Fritz Siebel for Figure 2.3, 'Weather Made to Order'; David Parkins for permission to use his cartoon, Figure 3.1, which first appeared in an issue of Nature in 2008; the IPCC for Figure 3.3, an image from their 5th Assessment Report; The National Research Council for Figure 3.4; Harvard University for the image from the SCoPEx website, Figure 3.5; Figure 4.1 reproduced with permission of the artist Shtig and www.geoengineeringmonitor.org; MIT Press for permission to reproduce the cover of David Keith's book as Figure 4.2; and the Royal Society for permission to produce Figure 5.1, from their 2009 assessment of geoengineering. Particular thanks to the artist Michael Portley for permission to use his painting 'Emissions Offset' for the cover.

Contents

1	**Introduction**	**1**
	The Basic Idea of Solar Geoengineering	5
	Imagining Solar Geoengineering	7
	Technoscience and the Sociotechnical	9
	Historical Precursors	11
	Knowledge-Power-Values	14
	'Official' Assessments and Knowledge-Brokers	16
	Outline of This Book	19
	References	22
2	**Geoengineering's Past: From Mastery to Taboo**	**27**
	Geoengineering's First Wave: Mastery	28
	Geoengineering's Hiatus	46
	Why Attempts to Revive SGE Failed	55
	Shifting Imaginaries of Climate, Nature and Globalisation	59
	References	68
3	**The Re-emergence of Solar Geoengineering**	**75**
	Climate Urgency and Emergency	76
	Geoengineering's Re-emergence	81

	Respectable but Not Embraced	89
	Rationales	96
	Frames and Metaphors	100
	Representations	105
	An Inability to Normalise	108
	References	112
4	**Competing Imaginaries of Solar Geoengineering**	123
	Un-Natural: The Perils and Injustice of Geopiracy	124
	Imperial: Acclimatising the World	129
	Chemtrails: Secret Elites Poisoning the World	140
	Convergences and Divergences in the Imaginaries	145
	A Sociotechnical Imaginary Struggling to Be Born	151
	The Imperial Imaginary's Quest for Legitimacy	154
	References	156
5	**Knowledge-Power-Values**	163
	Knowledge/s	164
	Power	179
	Values	194
	References	202
6	**Future Imaginings**	213
	The Climate of Power	214
	A Technology of the Powerful	217
	… And Also Disruptive	221
	Attaching SGE to More 'Optimistic' Imaginaries	223
	References	234
7	**Conclusion**	241
	The Argument	243
	What's Different?	244
	What's Constraining SGE?	246
	What's at Stake?	250

SGE's Shadow Presence in the Paris Agreement 256
Troubling Environmentalism 258
References 263

Index 267

About the Author

Jeremy Baskin is a Senior Research Fellow at the School of Government of the University of Melbourne in Australia. His current work focuses on geoengineering and climate policy, on the Anthropocene and global justice, and on the role of experts and expertise as it relates to environmental policy. Before entering academia Jeremy was an anti-apartheid activist, then a post-apartheid civil servant, before working globally with leaders in business, government, NGOs and multilateral agencies on the challenges of environmental sustainability.

Abbreviations

AEI	American Enterprise Institute
AR	Assessment Report (used in conjunction with IPCC)
BPC	Bipartisan Policy Center
CCC	Copenhagen Climate Consensus
CCS	Carbon Capture and Storage
CDR	Carbon Dioxide Removal
CO_2	Carbon dioxide (the most prevalent GHG)
CO_2e	Carbon dioxide equivalent (a measure of the warming potential of various GHGs if considered as an equivalent of CO_2)
CoP	Conference of the Parties (annual meeting on climate change under UNFCCC auspices)
ENMOD	United Nations Convention on the Prohibition of Military or Any Other Hostile Use of Environmental Modification Convention
GAO	United States Government Accountability Office
GeoMIP	The Geoengineering Model Intercomparison Project
GHG	Greenhouse Gas
GMO	Genetically Modified Organism
IPCC	Intergovernmental Panel on Climate Change
LLNL	Lawrence Livermore National Laboratory
NAS	National Academies of Sciences (United States)
NASA	National Aeronautics and Space Administration
NGO	Non-Governmental Organisation

NRC	National Research Council (United States)
ppm	parts per million
RCP	Representative Concentration Pathways (used by IPCC)
R2P	Responsibility to Protect
SAI	Sulphate Aerosol Injection
SGE	Solar geoengineering. Used throughout to describe the technique also known as Solar Radiation Management (SRM) by Sulphate Aerosol Injection (SAI)
SPICE	Stratospheric Particle Injection for Climate Engineering (project)
SRM	Solar Radiation Management
SRMGI	Solar Radiation Management Governance Initiative
STS	Science & Technology Studies
UNESCO	United Nations Educational, Scientific and Cultural Organisation
UNFCCC	United Nations Framework Convention on Climate Change
WG	Working Group (used in conjunction with IPCC)
W/m^2	Watts per square meter

List of Figures

Fig. 1.1	A standard depiction of geoengineering (Source: Lawrence Livermore National Laboratory)	10
Fig. 1.2	Summary periodisation of geoengineering context and associated imaginaries	20
Fig. 2.1	Summary periodisation of geoengineering context and associated imaginaries	28
Fig. 2.2	"Let's Remake Nature According to Stalin's Plan" (Artwork: Viktor Semenovich Ivanov, 1949)	32
Fig. 2.3	"Weather made to order" (Artwork by Fritz Siebel)	33
Fig. 3.1	Planet as patient, scientist as physician (Artwork with artist's permission: David Parkins, from *Nature*, (2008) *455*(7214), 737.)	102
Fig. 3.2	A standard depiction of geoengineering (Source: Lawrence Livermore National Laboratory)	105
Fig. 3.3	Representation of geoengineering in IPCC 5th Assessment Report (2013a: 632)	106
Fig. 3.4	Visualisations of SGE in NRC Report (2015: 33)	106
Fig. 3.5	Image from SCoPEx project website (Courtesy: Harvard University)	107
Fig. 4.1	Hands Off Mother Earth (HOME) sticker (Artwork: Shtig, courtesy of HOME Campaign)	128

Fig. 4.2 Book cover which encapsulate the 'Salvation' narrative (Courtesy: MIT Press) 141

Fig. 5.1 Royal Society summary evaluation of geoengineering techniques (2009: 49) 169

List of Tables

Table 2.1 Schematic summary of reigning geoengineering imaginaries
(1945–2005) 66
Table 3.1 Typology of arguments in favour of solar geoengineering 97
Table 4.1 Summary of competing imaginaries of solar geoengineering 146

1

Introduction

Can global warming be reversed or kept within 'safe' limits? Since at least the late 1980s the standard answer is that it can be … but only if greenhouse gas emissions are cut significantly, and soon. This approach informed the Kyoto treaty. And it is now enshrined in the Paris climate agreement. In reality though, climate action to date has been underwhelming. National emissions targets are insufficiently ambitious. The United States (and now Brazil) are no longer committed to the Paris agreement and its goals. Average global surface temperatures are currently about 1°C above pre-industrial levels and rising. So are greenhouse gas emissions. In the face of this gloomy outlook there is a different kind of solution circulating in climate policymaking circles, but still largely out of public view: that it is the earth, not human behaviour which should be modified. Specifically, the argument is being put forth by some that a cooler climate could be engineered.

One specific proposal, solar geoengineering (SGE),[1] has attracted substantial policy and research attention. The idea of SGE, simply put, is to

---

[1] It is an indication of the unsettled and controversial status of SGE that it goes by a variety of names. The term 'Solar Radiation Management' is widely used (for example IPCC 2013), although

inject sulphate particles into the stratosphere, perhaps a million or more tonnes per annum depending on the cooling effect being aimed for. The sulphate particles would circulate rapidly and envelop most of the globe. They would oxidise over a period of weeks to form a sulphuric acid aerosol which would reduce incoming sunlight by a small amount.[2] The idea is that by reflecting a portion of incoming sunlight before it enters the Earth system, global warming would be reduced or reversed, although to varying degrees in different locations. Apart from cooling, other physical climatic effects could include reduced rainfall, droughts and monsoon disruption, as well as damage to the ozone layer. But the extent of these likely effects, not to mention the associated social effects, is both contested and not reliably known.[3] Further, embarking on SGE is a long-term commitment, a multi-decadal commitment (at minimum) requiring stable and dependable global policy settings. Depending on the dosage, it is not easily terminated without serious climatic 'bounce-back' risk.

SGE is a radical and, on the face of it, thoroughly bad idea. It is regarded with incredulity and something approaching disgust by broad swathes of the public who have been asked their views (Buck 2018). In my assessment it is currently opposed by most climate scientists. Even the most prominent scientific proponent[4] of the idea, David Keith, acknowledges: "You

---

strictly speaking the sulphate aerosol approach is only one imagined technology for reducing incoming solar radiation, and 'management' suggests something far more administrative, practical and achievable than is likely the case. The terminology geoengineering by 'Sulphate Aerosol Injection' (SAI) is also widely used. The NRC report adopts the dull but descriptive label of "stratospheric aerosol albedo modification" (2015). All these terms refer to same technique, which I label 'solar geoengineering' (SGE). The more general term 'geoengineering' is similarly contested and is often called 'climate engineering'. Various IPCC reports use both (for example 2012). Efforts to put a positive gloss on the idea have suggested calling it "climate remediation" (BPC 2011), and terms such as "planet hacking" (Kintisch 2010) and "geopiracy" (ETC Group 2010), with clearly negative connotations, can also be found.

[2] Keith calculates that to counterbalance half of the current carbon dioxide forcing—that is, to offset about 120 billion tons of carbon—"… would require injecting only one million tons of sulphur into the stratosphere each year" (2013: 67).

[3] For a recent account of the science of SGE see Irvine et al. (2016).

[4] When I use the word 'proponent' it is important not to read 'cheerleader', but rather understand it as those who think SGE is a path down which we need to be willing to go, and for which we must actively prepare. Although some of SGE's proponents are enthusiastic, many are tentative and embrace it reluctantly as unavoidable or a 'lesser evil'. Tellingly, one of the leading participants in the United Kingdom's SPICE project calls his blog 'The Reluctant Geoengineer'.

are repulsed? Good. No one should like it. It's a terrible option" (quoted in Wagner and Weitzman 2012: n.p.). But he is not alone in arguing that, nevertheless, solar geoengineering is sadly necessary given the state of climate change. It might avert increasingly serious, perhaps even catastrophic, climate change, so the argument goes … or at least it might buy more time to adjust human behaviour and pursue a radically less emissions-intensive path. Other climate scientists, no less authoritative, argue that "the idea of 'fixing' the climate by hacking the Earth's reflection of sunlight is wildly, utterly, howlingly barking mad" (Pierrehumbert 2015: n.p.). And yet it is increasingly being taken seriously in high level policy circles, albeit in a dystopian register, presented as a 'lesser evil' or 'bad idea whose time has come'.

SGE entails shifting the boundaries at which human interventions in the more-than-human world are acceptable. But is also relates centrally to the question of who is authorised to shape this new world. Any adoption of SGE will mark a major step-change in climate policy and our climate future. Because SGE is not (yet) a deployed technology, how it is imagined and what stories accompany it are critical. Indeed, these will shape whether it becomes a deployed technology at all, because whilst SGE has entered into the mix of mainstream climate policy considerations, it is not yet officially embraced, and remains highly contentious.

The question then of why SGE is struggling to move from mainstream policy consideration to being fully embraced—or what is holding it back—is of interest to both proponents and opponents of SGE. This book is a critical exploration of this radical and highly contested idea. Given its magnitude, if it happens then it should happen with open eyes and with high levels of consent. It is my contention, and the focus of this book, that understanding how SGE is imagined and how it is socially located is key to understanding its social positioning at the threshold of acceptability, and what may become of it.

Drawing on Jasanoff's concept of 'sociotechnical imaginaries' (Jasanoff 2015) I identify three competing imaginaries of SGE—which I label an Imperial imaginary, an Un-Natural imaginary and a Chemtrail imaginary. None is hegemonic although the Imperial imaginary is preponderant amongst proponents. I analyse these imaginaries closely and explore their assumptions about science, expertise and what knowledge matters, about what values are important including the proper human relationship to 'nature', and about current relations of social and economic power in the world today. To simplify the argument, I suggest that if SGE is adopted with any legitimacy, it will be in large measure because its proponents have succeeded in presenting it as a mundane and unremarkable technology in our post-natural 'Anthropocene' world, and a facilitator of continuing development—in short, as a sociotechnical imaginary of the Anthropocene. Conversely, if SGE comes to be widely understood as a fundamentally elite project, and a highly risky and 'un-natural' one which exceeds the bounds of what humans are entitled to do, then it is less likely to be deployed. Another possibility is that powerful interests impose it regardless, but that course would have a very different valency to one which gains any kind of democratic imprimatur.

I will argue, contrary to the dominant view expressed in major assessments of SGE, that whether SGE is 'officially' embraced and deployed as a climate change solution, or indeed whether it is rejected and marginalised, has very little to do with conducting more scientific research aimed at reducing the many risks already identified and clarifying uncertainties. Even how global warming unfolds, whilst relevant, is of secondary importance.

Any examination of SGE soon makes it apparent that SGE is a 'troubling' technology in all senses of the word. It troubles, of course, the existing verities of climate policy—mitigation as primary solution to the warming problem—even when it is offered up as merely a supplement to mitigation. But it also troubles by bringing to the fore existential questions grounded in competing social values such as what it means to be human, as well as questions about the place of science in thinking about climate change, about 'progress' and about the human-nature relationship, and the geopolitics of global North and South. It impels reflection on environmentalism, democracy, how expertise is engaged in the shaping of policy, and much else besides. It shines a light on the dystopic dimensions of contemporary life in the West, and on the dark side of 'man's mastery of nature'. A focus on sociotechnical imaginaries allows us to pay more attention to such aspects than is common in 'scientific' analyses of SGE *as* technology.

The Basic Idea of Solar Geoengineering

For those less familiar with SGE it is worth getting a few basic facts onto the table, although as the book progresses it will become clear that even the most basic observations are not always straightforward.

Firstly, placing significant quantities of sulphate aerosols in the stratosphere is likely to induce planetary cooling, and rapidly. How reliably and predictably it would do so is far from clear.[5] Other physical effects SGE might have are contested. The highly influential US National Research Council (NRC) assessment of geoengineering lists some "consequences of concern" including ozone depletion, a "reduction in global precipitation", or rainfall, and that the climatic effects "will not be uniformly distributed around the globe" (2015: 7). This is not an exhaustive list (Robock 2008).

Secondly, the argument that SGE is relatively cheap (with costs purported to be in the low billions of dollars per annum), whilst contested, is plausible at least in comparison to the Paris solution and the associated cost of steep emissions cuts, major energy transitions and adaptations to a warming world. Of course 'cheap' can be a weasel word in the sense that a Bangladesh-made t-shirt is cheap only if one leaves out the social and environmental costs of its production. SGE's cheapness refers mainly to the direct costs of injecting the sulphate aerosols into the stratosphere using a fleet of aircraft.

Thirdly, SGE is being taken seriously but has not yet been officially embraced. Since the mid-2000s SGE has become part of mainstream climate policy discussions—not secretly, although largely out of the public eye. It has been proposed by a number of prominent climate scientists (see for example Crutzen 2006; Schneider 2001). It has been the object of analysis and assessment by leading scientific institutions in the USA, the UK, Germany and elsewhere (see for example Royal Society 2009; NRC 2015). It has been analysed and assessed by leading institutions, in the United States especially, which are close to the centres of political

---

[5] Major volcanic eruptions which shoot sulphate particles into the stratosphere which then spread globally are often seen as a natural analogue for SGE, a proof of concept.

power, such as the Bipartisan Policy Center (2011), NASA (Lane et al. 2007), the US Government Accountability Office (2011), the Council on Foreign Relations (Ricke et al. 2008), and the RAND Corporation (Lempert and Prosnitz 2011), amongst others. Research into geoengineering and SGE in particular has increased rapidly. Consideration of SGE now forms part of the work of the authoritative Intergovernmental Panel on Climate Change (see for example IPCC 2012) and there have been moves to add Geoengineering as a third leg of climate strategy alongside Mitigation and Adaptation.[6] It has not been 'officially' endorsed but neither has it been laid to rest.[7]

Fourthly, SGE is intended simply to mask global warming and does not address the physical drivers of warming. In the words of the major National Research Council (NRC) report entitled *Reflecting Sunlight to Cool the Earth*: "Albedo modification techniques mask the effects of greenhouse warming; they do not reduce greenhouse gas concentrations" (NRC 2015: 1). For this reason most, but not all, of SGE's proponents argue it should be deployed alongside continuing efforts to reduce GHG concentrations, either through emissions cuts and/or developing other geoengineering technologies of the Carbon Dioxide Removal (CDR) kind. I will be arguing that SGE masks much more than warming.

Finally, SGE is not yet an operating, deployed technology. It is currently being researched and imagined and some associated experimentation is underway. Jack Stilgoe has called it "… the figment of a particular technoscientific imagination" (2015: 6). But it is no less real for that, given the significant research energy dedicated to making it a reality, and the effect that even the idea of geoengineering has on the existing verities and assumptions of climate policy. SGE is both an emergent technology, or at least the idea of one, and a putative climate policy solution.

---

[6] Or a fourth leg if one considers Loss and Damage to be the third leg.
[7] The Paris climate agreement includes implicit backing for some geoengineering of the Negative Emissions Technology (NETs) variety, sometimes called Carbon Dioxide Removal (CDR). But it is silent about SGE, eerily so since, as I will discuss, SGE lurks in the text as a shadow presence. It is hard to see the temperature targets of 1.5°C or 2°C being met without utilising SGE (Xu et al. 2018).

Imagining Solar Geoengineering

How SGE is imagined is central to its unfolding and the focus of this book. It soon becomes apparent, on reading the geoengineering literature, that a range of visions and assumptions of how the world is and what it should be are projected onto SGE. What the climate condition is understood to be, what values are held dear, what the 'normal' state of the world is assumed to be and what world/s are aspired to, what is imagined to be possible and acceptable and what is not, all form part of the imagining of SGE. Which visions predominate will shape (or prevent) its emergence as a deployed technology.

When a leading proponent, David Keith, calls SGE "…a cheap tool that could green the world" (2013: x), a great many assumptions accompany this claim. So too when staunch opponents argue that geoengineering "would render the world dependent on technocratic elites, military-industrial complexes and transnational corporations to 'regulate' the global climate" (ETC Group/Biofuelwatch 2017).

The major institutional analyses and assessments of geoengineering (such as Royal Society 2009; IPCC 2012; NRC 2015), even as they acknowledge the relevance of the social, typically treat geoengineering as an object of scientific and technical expertise. That is, it is understood mainly as an exercise in shaping and re-shaping the physical climate system of the Earth. I turn this emphasis on its head and explore how SGE seems to act as a screen onto which competing visions of the world are projected. Imaginaries and the ideational are therefore central to my analysis. This is appropriate too, given that SGE is not yet an operational technology, and is currently more idea than operational 'thing', and that authority to transition from the former to the latter is at the heart of its contestation.[8] I identify three competing imaginaries which I label

---

[8] There are some similarities between this approach and that of science journalist Oliver Morton in his book *The Planet Remade* (2015). For Morton "the way a society imagines its future matters. Who gets to do the imagining matters" and utopian thinking is important, not because utopias are attainable, but because "imagining geoengineered worlds that might be good to live in, in which people could be safer and happier than they otherwise would be, is worth doing" (2015: 30). Morton's goal is what he calls "a deliberate planet", but one which people "take care not control" (2015: 344). I am less committed to validating the geoengineering turn.

Imperial, Un-Natural and Chemtrail. None is hegemonic although the Imperial imaginary is dominant amongst proponents of SGE and amongst reluctant endorsers in the climate science and climate policy communities.

Jasanoff's concept of 'sociotechnical imaginaries' is helpful here. In the introduction to her jointly edited collection *Dreamscapes of Modernity* (2015), she builds on the rich pre-existing vein of thinking about imaginaries (for example Anderson 1983; Taylor 2002) whilst trying to pay closer attention to "… modernity's two most salient forces: science and technology" (Jasanoff 2015: 6–7). For Jasanoff the concept of the 'sociotechnical imaginary' refers to:

> collectively held, institutionally stabilized, and publicly performed visions of desirable futures, animated by shared understandings of forms of social life and social order attainable through, and supportive of, advances in science and technology. (Jasanoff 2015: 3)

She notes that whilst "multiple imaginaries can coexist within a society", some are elevated to "… a dominant position for policy purposes". Further, "sociotechnical imaginaries can originate in the visions of single individuals, gaining traction through blatant exercises of power or sustained acts of coalition building" (2015: 3). Whilst sociotechnical imaginaries are "typically grounded in positive visions of social progress", there is an "interplay between positive and negative imaginings—between utopia and dystopia" (2015: 3). The resonance of the concept to analysis of SGE is apparent: utopia and dystopia do indeed co-exist, no single imaginary is yet dominant for policy purposes, and the 'Earth-shaping' and 'world-making' dimensions of geoengineering are entwined. All this impels us to understand the 'visions of desirable futures' which animate SGE.

I do, however, adapt the understanding of sociotechnical imaginary outlined above in one key respect. In my usage of the term I expand it to include not only how people imagine they fit with other people and with social structures, but also how they fit with the more-than-human world. It is hard to conceive a social order which lacks at least an implicit understanding of the environment: the 'natural' and material places within which that social existence is located. Geoengineering, of course, relates to activities at the intersection of the human and the natural world, and the perceived boundary (or its lack) between the social and the natural is central to this book.

Technoscience and the Sociotechnical

The major institutional assessments of geoengineering typically treat it largely as a technical and scientific challenge. In this they replicate the preponderant approach to global warming to be found in the work of the IPCC. Climate change is understood as primarily a scientific question. The analytical tools which are brought to bear are the standard ones of atmospheric physics and climate modelling (with assumptions about reduced incoming radiation added to existing General Circulation Models) alongside some use of a particular brand of economics able to generate commensurable data.

Two examples of this technoscientific approach to geoengineering serve to illustrate my point. Figure 1.1 is a typical example of the graphic representation of geoengineering and virtually identical representations appear in all the major assessment reports. No people or communities or eco-systems are imagined, only a range of engineered technologies to be applied. In similar vein the NRC report into geoengineering draws on standard mechanistic metaphors to explain the Earth's energy balance and its relationship to the climate problem.

> The climate system can be compared to a heating system with two knobs, either of which can be used to set the global mean temperature. The first knob is the concentration of greenhouse gases such as CO_2 in the atmosphere that affects the infrared side of the energy balance [and how much heat can escape into space] ... The other knob is the reflectance of the planet, which controls the amount of sunlight that the Earth absorbs. (NRC 2015: 33)

The assessments generally acknowledge that SGE raises major questions of governance, of conflicting values and of geo-politics. But these

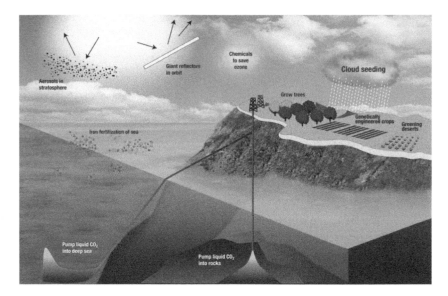

Fig. 1.1 A standard depiction of geoengineering (Source: Lawrence Livermore National Laboratory)

are not considered in any depth. Paradoxically, therefore, even as the centrality of values and visions is widely acknowledged to be central to thinking about geoengineering, the framing of the debate within a particular disciplinary hierarchy largely shuts this down, and proceeds to model options, costs, and effectiveness (in reducing temperature) in standard climate reductionist terms. In short, SGE is framed as a technoscientific issue.

Viewing SGE through the lens of technoscience conceals rather than illuminates what is at stake. As Szerszynski and Galarraga observe, "… the fashioning of climate [is] … a form of worldmaking" (2013: 2822–3). In this book I will be taking a different approach and emphasising the need to understand SGE as a sociotechnical project and not simply a technoscientific intervention. I embrace the insight from Science & Technology Studies (STS) that the social and the technological are always entwined (Bijker et al. 1987), and that science and society are co-produced (Jasanoff 2004). They come into being together and need to be thought together.

In some respects SGE can usefully be compared with other modern, 'big' technologies—such as nanotechnology, nuclear technology, and biotechnology—in that it raises comparably large existential and values questions. But it is also distinct. Whilst these technologies can be understood as innovative, and as having entailed scientific and technical advances, the same is not true of SGE. Technologically SGE is a fairly mundane combination of pre-existing technologies (balloons, aircraft, sprays, chemicals and so on). Scientifically it is largely derivative of existing knowledge in chemistry and the Earth sciences. But in its social and political aspects SGE is revolutionary and not mundane at all. Its aim to globally engineer key Earth systems makes it a totalising technology and this has vast social, political and ecological implications. Contrast too, for example, the pessimism which surrounds SGE with the techno-optimism attached at inception, at least in some countries, to nuclear power providing electricity 'too cheap to meter', or to the way in which biotech and nanotech have been able to associate themselves with a vision of 'progress', the opening of new horizons, the enhancement of personal choice, and the promise of plenty (Jasanoff and Kim 2009). SGE is different too in that whilst debate around similar technologies has often been rapidly shifted to technical and risk questions, debate around SGE remains open (Schäfer and Low 2014). Indeed, the centrality of the 'social' in assessing SGE is increasingly acknowledged although rarely practiced.

Historical Precursors

Understanding what is at stake, and the operative imaginaries, is enhanced by applying an historical lens. What we are witnessing today is not, strictly speaking, SGE's emergence, but its *re-emergence*. Ideas similar, sometimes even identical, to SGE have been in circulation, often at the highest political and military levels, since the conclusion of World War Two and for much of the Cold War. They have at times been embraced, and at other times been regarded as, in effect, taboo. By examining the sociotechnical imaginaries associated with previous incarnations of geoengineering, and both the continuities and discontinuities, we can illuminate both SGE's present and its possible futures.

The major institutional assessments of SGE typically avoid any reflection on this earlier history with its repeated displays of great hubris. Geoengineering discourse post-War was accompanied by both dreams and assumptions of Mastery and belief in the desirability and ability of the military-scientific-industrial elite to control and subdue nature (Fleming 2010). There is much to learn from that period. Although such unblushing hubris and contempt for the more-than-human world is out of fashion, doing SGE today still requires an assumption that it is possible to master the climate system and manage nature. But with the claim of mastery no longer easily available to prospective geoengineers (and experts generally), the utopian vision which once accompanied geoengineering is also no longer available either. Not surprisingly, therefore, contemporary SGE is widely understood as a dystopian project and this is a major reason why it struggles to gain a wider embrace.

An historical lens also helps one understand where the boundaries of what is acceptable and what is taboo have, or have not, shifted. There is an intriguing, but obscure, contemporary example which illustrates this: a proposal for a climate-focussed alternative to SGE other than emission reductions contained in a paper entitled 'Human Engineering and Climate Change'. It is important to note that this proposal has found no support from any institution, indeed what it proposes could be regarded as taboo. The authors note that responses to climate change have not been effective and that geoengineering is risky. They call instead for consideration of 'the biomedical modification of humans to make them better at mitigating climate change' (Liao et al. 2012: 207).[9] Interventions discussed include pharmacologically-induced meat intolerance, gene selection and hormone treatments to make new humans smaller (and thereby literally reduce their footprint), cognition enhancement aimed at reducing birthrates (smarter people apparently have fewer children), and pharmacological enhancement of altruism and empathy (since higher empathy levels are believed to correlate with better environmental behaviours). Liao, Sandberg and Roache argue, not unreasonably, that human engineering is "potentially less risky than geoengineering" and that it could make "behaviour and market solutions [to climate change] more likely to succeed" (2012: 211).

---

[9] Movie-goers may recognise some similarities with the storyline in the film *Downsizing*.

They analyse the objections that could be made to 'human enhancement' and respond to each one in detail. Many of these are remarkably similar to objections to SGE (including the charge of hubris and being 'un-natural'). But on the face of it they should be less troubling, especially given the human engineering they propose would be voluntary, albeit incentivised. Assuming it is not intended as parody, it is illuminating that Liao, Sandberg and Roache's proposal, to the best of my knowledge, has remained unloved. There have been no other supportive papers. There have been no feasibility assessments and no calls for further research.

Why is this proposal regarded as taboo whilst the taboo surrounding SGE is dissipating? Why is engineering the climate for humans thinkable, whilst engineering humans for climate is not? My purpose in posing these questions is to draw attention to three things to aid our exploration of the competing narratives of contemporary SGE. First, that the natural/unnatural binary remains a socially and historically powerful one, notwithstanding that it may rest on analytically shaky foundations. Second, that between taboo and normalisation lies an intermediate zone where a proposal is imagined to be possible, where it is no longer taboo but not yet normal. SGE is currently in this intermediate zone and engineering humans for climate change is not. Third, that whilst SGE is now treated as familiar, it is helpful to be able to retain a sense of its strangeness, as this is perhaps how most non-experts still regard SGE.

The ahistorical approaches found in most of the 'official' literature on geoengineering has practical implications in the present. For example, the assessments of SGE often suggest that contemporary volcanic eruptions should be closely monitored as they may be able to provide data useful in understanding the atmospheric effects of sulphates injected into the stratosphere. Data from the Pinatubo eruption of 1991 and more recent major eruptions is sometimes drawn upon. But older historical evidence is rarely mentioned. Examining major volcanic eruptions in history is hardly comforting and generally shows very strong climate perturbations and highly troubling societal effects. The massive Tambora eruption of 1815 resulted in plummeting global temperatures for a few years, failures of the Asian monsoon, major crop failures and accompanying famine and a global death toll possibly in the millions (Wood 2014). Critics of using historical evidence argue that SGE is different from such analogues both because modern society is believed to have greater resources for dealing with catastrophic social consequences and, on more technical grounds, including the spatial evenness of particle distribution under an effective spraying programme, and the relative cleanness of the stratosphere into which particles would be injected, compared to the lower atmosphere through which volcanic eruptions must pass (NRC 2015: appendix D). Others argue that the most recent modelling shows that if SGE was only used to offset half the anthropogenic warming there would be little change in global average rainfall and monsoon behaviour. The effect, however, is to gloss over evidence of actual, situated and socially-experienced climatic effects, in favour modelling climatic effects based on physical data alone and average global effects (Reynolds et al. 2016).

In short, for the range of reasons outlined above, I pay particular attention to earlier iterations of geoengineering and adopt a critical-historical perspective which illuminates the continuities and discontinuities in imaginaries of geoengineering.

Knowledge-Power-Values

How SGE is assessed is heavily dependent on what knowledge and what disciplines are relied upon, which value systems and normative goals are prioritised, and how the technology is understood to be connected to existing orderings of social, economic and geo-political power. The competing imaginaries of SGE that I identify—Imperial, Un-Natural and Chemtrail—stand differently in relation to knowledge, values and power. And yet, as I intend to show, the major institutional assessments of SGE all rely on approaches to knowledge, power and values which tend to reinforce the Imperial imaginary, even when they hold off from explicitly endorsing SGE.

In relation to knowledge I aim to explore which knowledge systems, technologies and epistemologies, and which 'epistemic communities' (Haas 1992) have been engaged with, and which have been ignored or relegated to a subsidiary role. I trace the implications of Szerszynski and Galarraga's insight that when it comes to geoengineering, the hierarchy of knowledge to date has prioritised the sciences, and the climate sciences in particular, with other disciplines clearly seen as unnecessary or supplemental (2013). My assumption is that it matters who shapes debates around geoengineering, who the key "knowledge-brokers" (Stone 2002) are, and what is assumed to count the most when assessing geoengineering.

In relation to values, I try to understand: what ontological orderings and values are contested in the process of SGE's re-emergence? I explore what assumptions, and what existential and normative concerns are present, and how the idea of SGE relates to some of the hegemonic ideas and value systems of our time, such as 'progress', 'development', 'the global', 'democracy' and human limits. SGE also shines a light on, and is a concentrated instance of, a major re-thinking of environmental politics that is currently underway. As a global intervention to moderate the planet's climate, it invites engagement with the wide variety of ontologies and cosmologies about 'Nature'. It compels reflection on whether we are now at the 'end of nature', whether 'nature' and its protection is a credible foundational concept for environmentalism, and the implications of the insight that "nature is always already social" (Swyngedouw 2011: 259). Contestation over SGE is intimately bound up with the widespread reimagining of environmental politics that has been variously labelled post-nature environmentalism, neo-green or eco-modernism (see for example Ecomodernist Manifesto 2015). As an extreme example of technological intervention into the Earth system, SGE shines a light on these new interpretations of environmentalism, and they, equally, may act to legitimate geoengineering (see for example Brand 2009). There is also, as I will show, some evidence that it is the perceived 'un-naturalness' of SGE, and a sense that it crosses a line of what it is 'proper' for humans to do, which are among the biggest obstacles to SGE's embrace.

In relation to power, my aim is to understand what material and geopolitical interests are served or might benefit from the adoption of SGE as a component of climate policy, and how power and the powerful are imagined in official and policy-influential accounts of SGE. In ambition SGE is itself clearly a powerful technology, one which aims to re-shape key components of Earth systems. I suggest it is also a technology of the socio-economically and geo-politically powerful, and explore what this means for the relationship between SGE and the political-economy of contemporary capitalism. Part of my enquiry therefore entails exploring the relationship between SGE and the dominant global economic order with its systems of extraction, production, consumption and trade. The Royal Society study, without endorsing its deployment, concluded that SGE, alone among the geoengineering technologies, could be assessed as both highly effective and highly affordable as well as fast-acting (2009). Not surprisingly, this makes solar geoengineering appealing to economic rationalists, potentially attractive to policymakers and seductive for politicians. At the same time, SGE's trans-boundary and uncontainable effects makes it hard to imagine how the proposed technology might be marketised or turned into an opportunity for profit—key drivers of any capitalist embrace.[10]

Central too in considering power, is the relationship between SGE and the existing geo-political order. Despite challenges to its hegemony, the United States remains the dominant global power and it is also the prime locus of consideration and research into SGE. One of the drivers that led to the commissioning of the NRC report into climate intervention (2015) was an anxiety by the US security establishment as to whether they would be able to detect if another country was embarking on geoengineering. Geoengineering, even as an idea and certainly if it becomes a practice, has power effects and implications. I explore how some of these effects may function to stabilise the dominant order, whilst others threaten to de-stabilise it. If SGE is compatible with capitalism, it is with an especially statist and imperial version of what Manfred Steger terms 'market globalism' (2009).

'Official' Assessments and Knowledge-Brokers

There is a large and rapidly growing literature on geoengineering. The reference list in the NRC report runs to 37 small-font pages (2015; see also Irvine et al. 2016), almost entirely from the peer-reviewed scientific

---

[10] This is not to suggest that there are no vested interests promoting geoengineering or likely to emerge, only that it is hard to see a for-profit industry interest at any great scale. Hamilton (2013) has pointed to the role of patents, ultra-rich benefactors (including Bill Gates), and the interests emerging from elements of the security establishment in promoting pro-geoengineering perspectives (see also Long and Scott 2013).

literature. There are now also hundreds of articles emanating from the social sciences and the humanities.

Because I am interested in the intersection of knowledge, power and values and how these shape imaginaries of geoengineering I have paid particular attention to what I call the 'institutional' or 'official' literature on SGE. These are reports, documents and assessments from authoritative institutions, or 'official' bodies in some sense. These are generally institutions which, domestically and internationally, influence the overall settings of climate policy: both current policies and potential future ones. They are typically close to one or more centres of institutional or governmental power and they accordingly help frame discourses of climate change. These reports are critical for understanding how a powerful technology is understood in the 'corridors of power'. Most commonly they are examples of what Jasanoff has called 'advisory science', a hybrid activity combining 'elements of scientific evidence with large doses of social and political judgement' (1990: 229). They have typically been produced by committees and then finessed and signed-off by the participants prior to release. As such they can be said to reflect some sort of collective, negotiated, perspective. In some instances they operate at the intersection of elite science, government, the policy elite, and even the military, with the NRC report (2015) being the most prominent example.

The most notable and influential institutional publications produced since SGE's re-emergence in the mid-2000s have been those emanating from the IPCC (especially associated with the Fifth Assessment Report), from the UK's Royal Society (2009), and the US National Research Council (2015). Relevant too are assessment reports commissioned by the US Government Accountability Office (2011), the Council on Foreign Relations (CFR), the RAND Corporation and the Bipartisan Policy Center (BPC). Their well-known closeness to the US bureaucratic, foreign policy, security and political establishments respectively gives them something verging on 'official' status, and at the very least an insight into thinking that is close to the US establishment. There have also been reports commissioned by the German Federal Ministry of Education and Research (Rickels et al. 2011) and the Umweltbundesamt (Ginzky et al. 2011), which adopt a more critical tone than their US counterparts, as well as collaborative projects such as the European Transdisciplinary Assessment of Climate Engineering (EuTRACE) (Schäfer et al. 2015). Finally, there are reports produced by coalitions of NGOs, such as Hands Off Mother Earth (HOME) which are overwhelmingly hostile to SGE (for example ETC/Biofuelwatch 2017) and provide insight into oppositional perspectives.

I have also paid close attention to what I term the 'policy-influential' literature. These are works—journalistic, popular and academic—by individual authors, often but not always scientists, who have been closely associated with a number of the institutional reports. These are the key knowledge-brokers and scene-setters of SGE. They have generally played a central role in framing the institutional debates around geoengineering and presenting the ideas to politicians and the public. They frequently assess specific geoengineering proposals, recommend funding, and occasionally advise ultra-wealthy individuals with an interest in geoengineering (Hamilton 2013: 72ff). They often referee reports where they have not been panellists, and are repeatedly cited in press articles on the topic, thereby helping frame opinion. Their writings are less constrained than the 'institutional' literature, and more likely to engage with the implications of geoengineering, or address extra-scientific issues that are causing disquiet. Their views influence climate policy.[11]

Throughout, I have stood on the shoulders of others and have tried to engage with their work. In 2013 Harvard University's David Keith, the most active of the knowledge-brokers, published a short monograph entitled *A case for climate engineering*, specifically a case for solar geoengineering. This elicited a sharp response from Mike Hulme, a well-known UK-based geographer in the form of a book entitled *Can Science Fix Climate Change? A case against climate engineering* (2014). Hulme takes issue with hubristic techno-fixes and argues that climate engineering is

---

[11] Those generally supportive of geoengineering have been labelled the 'geo-clique'. It is a term which has gained traction but which I avoid. Nevertheless, it is the case that there is a small, but expanding, and relatively close-knit group of individual knowledge-brokers who play an outsized role in shaping the debate around SGE. Key names commonly identified include Ken Caldeira, John Shepherd, David Keith, David Victor, Jason Blackstock, Ben Kravitz, Peter Irvine, Andy Jones, Philip Rasch, Andy Parker, Scott Barrett, Gernot Wagner and Michael MacCracken (Kintisch 2010: 8; Oldham et al. 2014)—brokers at the centre of debates around geoengineering (Möller n.d., unpublished). It may or may not be relevant that the key knowledge-brokers are overwhelmingly pale, male, First-world scientists.

'undesirable, ungovernable and unreliable' (2014: xii). Both books are important exemplars of strong, contrasting perspectives on SGE emanating from the sciences and social sciences respectively.

Other literature which has been influential includes edited collections focussing on the scientific and engineering issues associated with different geoengineering techniques (for example Launder and Thompson 2010; Harrison and Hester 2014); reflections on the ethics and governance and hubris of geoengineering (such as Preston 2016; Blackstock et al. 2015; Gardiner 2011; Hamilton 2013); anecdote-rich journalistic accounts (Kintisch 2010; Goodell 2010); studies locating geoengineering historically such as Jim Fleming's *Fixing the Sky* (2010); and policy influential work such as David Victor's *Global Warming Gridlock* (2011), which argues that in a "climate emergency a well-designed geoengineering plan will be better than doing nothing" (2011: 166). Jack Stilgoe's *Experiment Earth* (2015) examines solar geoengineering, including the Stratospheric Particle Injection for Climate Engineering (SPICE) project, and is inspired by the sociology of expectations. His characterization of solar geoengineering as "the figment of a particular technoscientific imagination" has much in common with my own approach. But where he is concerned with how good and democratic experimentation might emerge, my focus is on the competing imaginaries themselves and their relationship to social order.

Outline of This Book

The Chapters between this Introduction and the book's Conclusion broadly cover SGE's Yesterday (Chapter 2), Today (Chapters 3, 4 and 5), and Tomorrow (Chapter 6).

Chapter 2 explores the history and understandings of geoengineering in the period prior to 2005 and draws out the continuities and discontinuities between how geoengineering, and especially SGE, have been understood and imagined at different times. I identify three phases in geoengineering's history prior to its recent re-emergence: Mastery, Unimaginability and Taboo. Figure 1.2 depicts this periodisation and the associated contexts and imaginaries of the times. I track the enthusiastic

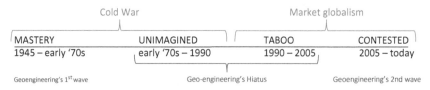

Fig. 1.2 Summary periodisation of geoengineering context and associated imaginaries

embrace of geoengineering as an idea after 1945 and show how this formed part of a larger Cold War imaginary. In the West this involved the triad of modernity: a belief in progress, the invention of development, and a faith in science and the technologies it could generate, and with all three underpinned by an assumption of human entitlement, even obligation, to mastery over nature. These had almost identical counterparts in the Soviet sphere. Geoengineering was embraced because it was possible, and as part of US-Soviet Cold War techno-rivalry, and not because there was a climate problem needing a solution. I go on to examine the hiatus period, an interregnum in which geoengineering was seen as unnecessary and unwise and was no longer commonly imagined, before coming to be regarded as taboo. It was a taboo that persisted even during an unsuccessful attempt by some US scientists to revive the idea in the early 1990s. This seeks to understand the reasons for geoengineering's effective disappearance from the climate policy table, and its relationship to the rise of neo-liberal globalisation from the 1980s. It reveals some of the specific thinking about science, climate and nature that made geoengineering essentially (although not literally) unimaginable.

Chapter 3, 'The Re-emergence of Solar Geoengineering', charts the re-emergence of SGE into mainstream consideration as a policy option from the mid-2000s, linking this to growing concerns about climate. I analyse the institutional literature and assessments in depth, and trace the most common rationales that accompanied SGE's revival, and analyse the various ways in which it is framed, understood and imagined today. I show how geoengineering has been unable to be normalised as a respectable third leg of climate policy. It is researched and imagined but it lacks legitimacy and traction.

In Chapter 4 I analyse three competing imaginaries of SGE and their associated logics, assumptions and narratives. I identify the most prominent as the Imperial imaginary which brings together three narrative strands ('Market', 'Geo-management' and 'Salvation'). I also identify two oppositional imaginaries, one which I label the Un-Natural imaginary, and another which I label the Chemtrail imaginary and which operates in a different register. I examine the implicit and explicit values, worldviews and ontological orderings in these competing imaginaries. Each of these engages with SGE beyond a purely technoscientific framing, but none has yet managed to emerge as hegemonic. Without a hegemonic narrative it becomes difficult for SGE to be officially embraced and normalised as both applied technology and climate policy.

In Chapter 5, 'Knowledge-Power-Values', I dive more deeply into understandings of SGE today, in an effort to understand attempts to stabilise and normalise it as a concept. I focus on epistemological and knowledge choices, questions of values and ontological ordering, and the relationship of SGE to power and the powerful. In particular I cast a critical eye on the elevation of the "techno-scientific" in institutional assessments of SGE, the invocation of 'emergency', the elevation of cost-benefit thinking, the narrowing of ethical debates, and the implications of SGE for capitalism and geo-political ordering. In each case I show how contradictory forces and imperatives cohabit, in ways which restrain normalisation.

Chapter 6 is future focussed and therefore more speculative. I look in particular at three dimensions which I expect to be relevant in shaping whether SGE is embraced and deployed, or spurned and discarded. These are: the relevance of actual changes in climate and weather; the extent to which elites see SGE as essential to stabilising the dominant geo-political and socioeconomic order; and the degree to which SGE's proponents are able to replace the predominantly dystopian vision which currently accompanies it. In relation to the last of these, I explore how paradigms of 'development' and the Anthropocene have begun to be mobilised in the reframing of SGE as a positive project of modernity.

Finally, the concluding Chapter draws the key strands of the argument together and reflects on what's different about SGE past and present, what's constraining its normalisation, and what's at stake in any decision to embrace it? In the process I look at a number of the implications of my analysis for thinking about climate policy post-Paris and about environmentalism more generally, before considering whether SGE will ever come into being as a deployed technology and a third leg of climate policy. I hope it will not, but I fear that it might.

References

Anderson, B. (1983). *Imagined communities*. London: Verso.
Bijker, W., Hughes, T. P., & Pinch, T. (Eds.). (1987). *The social construction of technological systems: New directions in the sociology and history of technology*. Cambridge, MA: MIT Press.
Bipartisan Policy Center (BPC). (2011). *Geoengineering: A national strategic plan for research on the potential effectiveness, feasibility, and consequences of climate remediation technologies*. Washington, DC: Bipartisan Policy Center Task Force on Climate Remediation Research.
Blackstock, J., Miller, C., & Rayner, S. (Eds.). (2015). *Geoengineering our climate? Ethics, politics and governance*. London: Routledge.
Brand, S. (2009). *Whole earth discipline: An ecopragmatist manifesto*. London: Viking.
Buck, H. J. (2018). Perspectives on solar geoengineering from Finnish Lapland: Local insights on the global imaginary of Arctic geoengineering. *Geoforum, 91*, 78–86.
Council on Foreign Relations (CFR). (2008). See Ricke et al. 2008.
Crutzen, P. J. (2006). Albedo enhancement by stratospheric sulfur injections: A contribution to resolve a policy dilemma? *Climatic Change, 77*(3–4), 211–219.
Ecomodernist Manifesto. (2015). Retrieved January 9, 2019, from http://www.ecomodernism.org/
ETC Group. (2010, November). *Geopiracy: The case against geoengineering* (2nd ed.). Retrieved January 9, 2019, from https://www.cbd.int/doc/emerging-issues/etcgroup-geopiracy-2011-013-en.pdf
ETC Group/Biofuelwatch. (2017). *The big bad fix: The case against climate geoengineering*. Retrieved January 9, 2019, from http://etcgroup.org/sites/www.etcgroup.org/files/files/etc_bbf_mar2018_us_v1_web.pdf
Fleming, J. R. (2010). *Fixing the sky: The checkered history of weather and climate control*. New York: Columbia University Press.

GAO (Government Accountability Office). (2011). *Climate engineering: Technical status, future directions, and potential responses.* GAO-11-71. Washington, DC: U.S. Government Accountability Office. Retrieved January 9, 2019, from http://www.gao.gov/new.items/d1171.pdf

Gardiner, S. (2011). *A perfect moral storm: The ethical tragedy of climate change.* Oxford: Oxford University Press.

Ginzky, H., Herrmann, F., Kartschall, K., Leujak, W., Lipsius, K., Mäder, C., et al. (2011). *Geoengineering: Effective climate protection or megalomania?* Dessau-Roßlau: Umweltbundesamt.

Goodell, J. (2010). *How to cool the planet: Geoengineering and the audacious quest to fix Earth's climate.* Melbourne: Scribe.

Haas, P. M. (1992). Introduction: Epistemic communities and international policy coordination. *International Organization, 46*, 1–35.

Hamilton, C. (2013). *Earthmasters: Playing god with the climate.* Crow's Nest, NSW: Allen & Unwin.

Harrison, R. M., & Hester, R. E. (Eds.). (2014). *Geoengineering of the climate system.* Cambridge: Royal Society of Chemistry.

Hulme, M. (2014). *Can science fix climate change? A case against climate engineering.* Cambridge, UK: Polity Press.

IPCC (Intergovernmental Panel on Climate Change). (2012). *Meeting report of the Intergovernmental Panel on Climate Change expert meeting on geoengineering* (O. Edenhofer, R. Pichs-Madruga, Y. Sokona, C. Field, V. Barros, T. F. Stocker, Q. Dahe, J. Minx, K. Mach, G.-K. Plattner, S. Schlömer, G. Hansen, & M. Mastrandrea, Eds.). IPCC Working Group III Technical Support Unit, Potsdam Institute for Climate Impact Research. Geneva: IPCC.

IPCC (Intergovernmental Panel on Climate Change). (2013). Summary for policymakers. In T. F. Stocker, D. Qin, G.-K. Plattner, M. Tignor, S. K. Allen, J. Boschung, A. Nauels, Y. Xia, V. Bex, & P. M. Midgley (Eds.), *Climate Change 2013: The physical science basis. Contribution of Working Group I to the Fifth Assessment Report of the Intergovernmental Panel on Climate Change.* Cambridge: Cambridge University Press.

Irvine, P. J., Kravitz, B., Lawrence, M. G., & Muri, H. (2016). An overview of the Earth system science of solar geoengineering. *WIREs Climate Change, 7*(6), 815–833.

Jasanoff, S. (1990). *The fifth branch: Science advisers as policymakers.* Cambridge, MA: Harvard University Press.

Jasanoff, S. (2004). The idiom of co-production. In S. Jasanoff (Ed.), *States of knowledge: The co-production of science and social order.* London and New York: Routledge.

Jasanoff, S. (2015). Future imperfect: Science, technology and the imaginations of modernity. In S. Jasanoff & S.-H. Kim (Eds.), *Dreamscapes of modernity: Sociotechnical imaginaries and the fabrication of power* (pp. 1–33). Chicago: Chicago University Press.

Jasanoff, S., & Kim, S.-H. (2009). Containing the atom: Sociotechnical imaginaries and nuclear power in the United States and South Korea. *Minerva, 47*(2), 119–146.

Keith, D. W. (2013). *A case for climate engineering.* Cambridge, MA: MIT Press.

Kintisch, E. (2010). *Hack the planet: Science's best hope—Or worst nightmare—For averting climate catastrophe.* Hoboken, NJ: Wiley.

Lane, L., Caldeira, K., Chatfield, R., & Langhoff, S. (2007). *Workshop report on managing solar radiation.* NASA Ames Research Centre & Carnegie Institute of Washington, Moffett Field, CA, 18–19 November. Hanover, MD: NASA. (NASA/CP-2007-214558) Retrieved from http://event.arc.nasa.gov/main/home/reports/SolarRadiationCP.pdf

Launder, B., & Thompson, J. M. T. (Eds.). (2010). *Geo-engineering climate change: Environmental necessity or Pandora's box?* Cambridge: Cambridge University Press.

Lempert, R. J., & Prosnitz, D. (2011). *Governing geoengineering research: A political and technical vulnerability analysis of potential near-term options.* Santa Monica, CA: RAND Corporation.

Liao, S. M., Sandberg, A., & Roache, R. (2012). Human engineering and climate change. *Ethics, Policy & Environment, 15*(2), 206–221.

Long, J. C. S., & Scott, D. (2013). Vested interests and geoengineering research. *Issues in Science and Technology, 29*(3), 45–52.

Möller, I. (n.d.). The geoengineering knowledge network: Explaining norm convergence on geoengineering in global climate politics. Unpublished paper supplied by author.

Morton, O. (2015). *The planet remade: How geoengineering could change the world.* London: Granta.

NASA. 2007. See Lane et al. 2007.

National Research Council (NRC). (2015). *Climate intervention: Reflecting sunlight to cool Earth.* Washington, DC: National Academy of Sciences.

Oldham, P., Szerszynski, B., Stilgoe, J., Brown, C., Eacott, B., & Yuille, A. (2014). Mapping the landscape of climate engineering. *Philosophical Transactions of the Royal Society A, 372*(2031), 1–20.

Pierrehumbert, R. (2015, February 10). Climate hacking is barking mad. *Slate.* Retrieved January 9, 2019, from http://www.slate.com/articles/health_and_

science/science/2015/02/nrc_geoengineering_report_climate_hacking_is_dangerous_and_barking_mad.html

Preston, C. (Ed.). (2016). *Climate justice and geoengineering ethics and policy in the atmospheric Anthropocene*. London: Rowman & Littlefield.

RAND Corporation. (2011). See Lempert & Prosnitz 2011.

Reynolds, J. L., Parker, A., & Irvine, P. (2016). Five solar geoengineering tropes that have outstayed their welcome. *Earth's Future, 4*(12), 562–568.

Ricke, K., Morgan, M. G., Apt, J., Victor, D., & Steinbruner, J. (2008, May 5). *Unilateral geoengineering: Non-technical briefing notes for a workshop at the Council on Foreign Relations*. Washington, DC. Retrieved January 9, 2019, from http://www.cfr.org/content/thinktank/GeoEng_Jan2709.pdf

Rickels, W., Klepper, G., Dovern, J., Betz, G., Brachatzek, N., Cacean, S., et al. (2011). *Large-scale intentional interventions into the climate system? Assessing the climate engineering debate*. Scoping report conducted on behalf of the German Federal Ministry of Education and Research (BMBF), Kiel Earth Institute.

Robock, A. (2008). 20 reasons why geoengineering may be a bad idea. *Bulletin of the Atomic Scientists, 64*(2), 14–18.

Royal Society. (2009). *Geoengineering the climate: Science, governance and uncertainty*. RS Policy document 10/09. London: Royal Society. Retrieved January 9, 2019, from https://royalsociety.org/~/media/Royal_Society_Content/policy/publications/2009/8693.pdf

Schäfer, S., Lawrence, M., Stelzer, H., Born, W., & Low, S. (2015). *Removing greenhouse gases from the atmosphere and reflecting sunlight away from Earth*. Final report of the FP7 CSA project EuTRACE (European Transdisciplinary Assessment of Climate Engineering). Potsdam: Institute for Advanced Sustainability Studies.

Schäfer, S., & Low, S. (2014). Asilomar moments: Formative framings in recombinant DNA and solar climate engineering research. *Philosophical Transactions of the Royal Society A, 372*(2031), 1–15.

Schneider, S. (2001). Earth systems engineering and management. *Nature, 409*(6818), 417–421.

Steger, M. B. (2009). *Globalisms: The great ideological struggle of the twenty-first century* (3rd ed.). Lanham, MD: Rowman & Littlefield.

Stilgoe, J. (2015). *Experiment Earth: Responsible innovation in geoengineering*. Abingdon: Routledge.

Stone, D. (2002). Introduction: Global knowledge and advocacy networks. *Global Networks, 2*(1), 1–11.

Swyngedouw, E. (2011). Depoliticized environments: The end of nature, climate change and the post-political condition. *Royal Institute of Philosophy Supplement, 69,* 253–274.

Szerszynski, B., & Galarraga, M. (2013). Geoengineering knowledge: Interdisciplinarity and the shaping of climate engineering research. *Environment and Planning A, 45*(12), 2817–2824.

Taylor, C. (2002). Modern social imaginaries. *Public Culture, 14*(1), 91–124.

Victor, D. G. (2011). *Global warming gridlock: Creating more effective strategies for protecting the planet.* Cambridge: Cambridge University Press.

Wagner, G., & Weitzman, M. L. (2012, October 24). Playing God. *Foreign Policy.* Retrieved January 9, 2019, from http://foreignpolicy.com/2012/10/24/playing-god/

Wood, G. D. (2014). *Tambora: The eruption that changed the world.* Princeton, NJ: Princeton University Press.

Xu, Y., Ramanathan, V., & Victor, D. G. (2018). Global warming will happen faster than we think. *Nature, 564,* 30–32.

2

Geoengineering's Past: From Mastery to Taboo

Solar geoengineering (SGE) emerged as a climate policy option from the mid-2000s. But strictly speaking, it *re-emerged* from the closet. There were earlier efforts, after the Second World War, to pursue geoengineering in general and solar geoengineering in particular. Understanding the imaginaries and practices that animated that First Wave, the reasons it failed then, and the continuities and discontinuities between the past and the present, can help us to make sense of solar geoengineering today.

In this Chapter I explore the history and understandings of geoengineering in the period up until the early 2000s. Figure 2.1 summarises the periodization I adopt. The First Wave emerged in the wake of the Second World War, ran concurrently with the first period of the Cold War, and could be found in both the East and the West. Lasting for almost three decades (from 1945 until the early 1970s), this phase was grounded in a manifest confidence in the idea that 'we can and should' geoengineer. I label this First Wave of geoengineering the Mastery phase, since it was characterised by dreams and assumptions of mastery. From the early 1970s geoengineering enters a Hiatus. This comprises two stages. The first, starting in the early 1970s, sees scientific and policy interest in climate engineering rapidly dissipate and the idea largely disappears from sight and becomes Unimagined. At a time when global

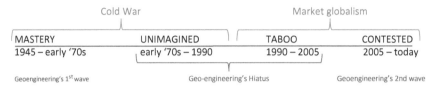

Fig. 2.1 Summary periodisation of geoengineering context and associated imaginaries

warming is increasingly understood to be anthropogenic, mitigation is assumed to be the required response, not geoengineering. The second stage of the Hiatus period, from shortly after the end of the Cold War until the early 2000s, sees unsuccessful attempts to revive interest in geoengineering as a policy option. Its proponents find that geoengineering is widely regarded as unacceptable and as, in effect, Taboo, something not to be countenanced in polite company.

Geoengineering's First Wave: Mastery

The beginnings of the Cold War can be found in the closing months of the Second World War. In Europe there was a race between the nominal allies (mainly Britain, the United States and the Soviet Union) to ensure that Hitler's defeat left them in control of the largest possible swathes of territory. Churchill (and De Gaulle's Free French) were committed to regaining control of those colonial possessions they had been forced to abandon, and re-asserting their diluted authority over those they had retained throughout the War. National resistance movements, of various stripes but generally communist in orientation, were advancing in a range of places including China, South-East Asia, Korea, Indonesia and elsewhere.

There was concern in Washington about future United States (US) influence in the Pacific and a desire to conclude the war there on favourable terms. Fearful of the possibility of the Soviet Union and Japan concluding a separate peace, and wanting to set down a marker for the future, the US decided to deploy its newly-invented atomic bomb to force Japanese surrender. The Soviets, whose infrastructure and population had been hardest hit by the war, would soon devote vast resources to science and military technology to ensure they would not be left behind. They emerged from the War as the second global superpower, after the United States. Competition between them was central to what would soon become known as the 'Cold War'. Ideology played a key role in shaping this competition and how it played out globally. America, as Westad puts it, cast itself as an 'empire of liberty', whilst the Soviet Union cast itself an 'empire of justice', and both thought of themselves as the true bearers of Western modernity (2007).

Over the following decades, the United States developed its military power and global presence, whilst the Soviet Union attempted to keep up with it and, occasionally, surpassed it. The connection between scientific production, government funding, and military application was generally substantial throughout the Cold War and was integral to the arms race (Oreskes and Krige 2014; Masco 2010). The period saw enormous investment, in both the US and the USSR, aimed at advancing both climate control capability and the science of meteorology more generally, with both believed to be centrally important to the military and to their nuclear weapons programmes in particular. Fleming (2010a) has labelled the Cold War proponents of climate engineering as the "weather warriors", since their scientific work, proposals and even their fantasies were deeply enmeshed with the military.

In the United States, perhaps because of the role of that country's scientific and technological achievements in winning the war, what emerged was a belief that almost any problem, natural or man-made, could be fixed. Almost identical views could be found in the Soviet Union, underpinned in that case by the Stalinist belief in the power of the socialist order to overcome all obstacles to 'progress'. In hindsight what is most striking is that, after the most destructive and technologically sophisticated war (and genocide) in history, both East and West embraced a highly hubristic vision of 'progress', science and technology. Indeed, this seemed to be a central component of the early Cold War *zeitgeist*.

30   J. Baskin

Technophilia, Hubris and the Early Cold War

Julian Huxley, a well-known British physicist and self-declared humanist (and soon to be founding head of UNESCO), flew to New York in December 1945, in the aftermath of the bombing of Hiroshima and Nagasaki. According to a *New York Times* report on the 20,000-strong gathering he addressed, he called for international control of atomic energy, something then being strongly resisted by the US military. Better, argued Huxley, to use atomic power to help "regions which are still no better off than they were a century and more ago." It could be used as "atomic dynamite" for "landscaping the Earth". 'Dams could be built in a fraction of the time'. Arctic pack ice and snow could be melted.

When challenged about the Arctic proposal Huxley admitted, said the reporter, that the exact consequences were unclear. 'He would not blow the Arctic cap away all at once ... A beginning would be made with "only a little bit of ice, north of Labrador, say"'. Huxley was equally enthusiastic about atomic power. It might distil seawater and flood the Sahara so that it might "again blossom". 'Atomic power could make an old dream come true', 'opening regions which are now uninhabitable or economically uninteresting', and be 'a safeguard against the social danger of overdeveloping densely populated industrial regions'. Huxley's vision, said the reporter, involved 'international control of atomic energy and ... social planning on a world-wide scale' (New York Times 1945, December 9).

I cite Huxley's stance at some length[1] because his views capture both technological hubris and the desire to impose a certain type of developmentalism upon the world—literally blasting the backward and 'uninteresting' regions into modernity. This was before the term 'development', as we use it today, had been invented (Sachs 2010).

Highly influential US public intellectual Arthur Schlesinger Jnr., remarked in his 1949 book *The Vital Center: The Politics of Freedom*, that:

---

[1] Huxley's own biography is fascinating with a combination of interests virtually unthinkable today. He was an influential scientist, a committed eugenicist, a colonialist, an old-style conservationist (and founder of WWF), and first head of UNESCO (although he was ousted before his term ended, being perceived as too left-wing once the Cold War commenced in earnest).

… no people in the world approach the Americans in mastery of the new magic of science and technology. Our engineers can transform arid plains or poverty stricken river valleys into wonderlands of vegetation and power. (cited in Bonnheim 2010: 892)

Similar thinking could be found in the Soviet Union. In 1948 Stalin had announced his plan for 'The Great Transformation of Nature', what Brain has called 'the world's first state-directed effort to reverse human-induced climate change' (2010: 671). Shaw cites the editors of *Voprosy Geografii*, published in 1952 to address *The Tasks of Physical Geography in Connection with the Great Construction Works of Communism*: "Whilst remaining natural scientists, geographers must at the same time become in their own way 'builders' of new, transformed landscapes, or 'engineers of nature'" (Shaw 2015: 127).

Arkadii Borosovich Markin's 1956 book *Soviet Electric Power* emphasised the imagined future role of nuclear power:

New islands and colossal dams will be built and new mountain chains will appear. Atom explosions will cut new canyons through mountain ranges and will speedily create canals, reservoirs, and sea, carry[ing] out huge excavation jobs…. Science will find a method of protection against the radiation of radioactive substances. (cited in Fleming 2010a: 199)

These perspectives were a continuation of longstanding Soviet interest in subjugating nature, and opening up the Arctic and Siberia in particular, with science and technology seen as the key to this (Josephson 2011: 420). They had almost identical counterparts in the US government's 'Operation Plowshare' initiative of the 1950s, a geological engineering programme to "… research, design, and implement the 'constructive' uses of nuclear explosives" with its initial focus on opening up Alaska (Kirsch and Mitchell 1998: 101).

Igor Adabashev, in his 1966 book *Global Engineering*, also echoed some of Huxley's developmentalist dreams but he attached them to the Soviet self-identification as bearer of justice. He envisaged diverting rivers and creating a vast inland sea in the Sahara, thereby allowing crop production to flourish. This would enhance 'the struggle of African

Fig. 2.2 "Let's Remake Nature According to Stalin's Plan" (Artwork: Viktor Semenovich Ivanov, 1949)

people's for national liberation'. Such projects had been delayed by capitalism which was like 'a ball and chain hampering man in his progress towards a happier lot' (cited in Fleming 2010a: 200).

Propagandist and popular imagery also convey a sense of the times. Figure 2.2 reproduces a 1949 Soviet poster. This portrays Stalin as fatherly leader explaining his plan for the 'Transformation of Nature' to two heroic-looking workers. Behind Stalin, the 'forest breaks' (like wooded barriers) described in the plan and surrounding the collective farm fields, are already in place. The slogan reads "Let's Remake Nature According to Stalin's Plan".

Figure 2.3 comes from a 1954 cover of a popular US magazine, *Collier's*. It is suggestive of popular perceptions at that time regarding the promise of engineering the weather. It also directly alludes to the military interest in this.

In short, from the outset, on both sides of the Cold War, the dominant imaginary was one of Mastery. It involved overcoming Nature, in the name of 'progress' and 'development', by mobilising science and technology, and in the most hubristic combination imaginable. It was an overwhelmingly optimistic vision. The technoscience imagined was state-driven, envisaged only 'big' technologies, and the purportedly 'socialist'

2 Geoengineering's Past: From Mastery to Taboo 33

Fig. 2.3 "Weather made to order" (Artwork by Fritz Siebel)

and 'capitalist' (or 'justice' and 'liberty') visions of what 'progress' and 'development' entailed sounded remarkably similar.

Edward Teller, a leading US scientist, and an influential figure in both the technosciences and politics of the era (and, later, in solar geoengineering), captured the mood of the time in his 1962 book *The Legacy of Hiroshima*.[2] Some typical quotations from that book:

---

[2] Teller was a key figure in the practice and politics of 'big science' throughout the Cold War and beyond. US President Eisenhower, in his 1961 farewell speech, famously warned against both the

We would be unfaithful to the tradition of Western Civilization if we shied away from exploring what man can accomplish, if we failed to increase man's control over nature … (p. 56).

… the demands of the world's needy must be met and can be met, and they will be met through science (p. 81).

Science brings progress; progress creates power; power is coupled with responsibility (p. 93).

The ancient story of Prometheus has ceased to be a legend. It has become a fact. There are among us those who are frightened by such progress, those who would turn back. This we cannot do (p. 299). (Teller 1962)

Geological Engineering and Climate Modification

This vision of Mastery was widely held not only by the political and military elites, but also by leading scientists and large swathes of the scientific community. Of relevance to climate modification and the genealogy of geoengineering, three related practices stood out: geological engineering, climate/weather control (the distinction was blurred at this time), and the development of meteorology (and associated computing, rocket/satellite and space capabilities) and cognate disciplines such as oceanography.

It is important to note that for much of this time whilst the warming physics of CO_2 concentrations had been theorised, the trend of actual warming was not yet known nor was 'climate change' a framework of understanding. Hence, the fantasies and proposals covered both weather control and climate control, and the boundaries between these objectives were never clear. Indeed, says Miller, in this period "… climate and weather … were essentially identical" (2004: 50). Because solar geoengineering is our focus, the account below emphasises proposals aimed at atmospheric engineering, although in practice these were not always clearly distinguished from other interventions.

In October 1945, two months after the dropping of atomic bombs on Japan, Vladimir Zworykin's "Outline of a Weather Proposal" raised the

'military-industrial complex' and the hijacking of public policy by the 'scientific-technological elite', and he let it be known afterwards that it was Teller and Werner von Braun that he had in mind (Goodchild 2004: xxiii).

need for more exact, scientifically-based weather knowledge. Better data, and the ability to compute, might allow for effective weather control. According to Jim Fleming, a leading historian of this period, Zworykin imagined a future where "[l]ong-term climatic changes might be engineered by large-scale geographical modification projects…" (2010a: 192–3). Zworykin, a research director at the RCA Laboratory in Princeton, imagined an international organisation to both study and channel the weather and thereby 'improve' the climate, in the hope that its "… eventual far-reaching beneficial effects on the world economy may contribute to the cause of peace" (cited in Fleming 2010a: 193).

Zworykin's proposal was immediately endorsed by the influential John von Neumann, who had been central to developing the atomic bomb, even helping select the Japanese cities to be targeted (Macrae 1992).[3] In von Neumann's endorsement he envisioned "altering the reflective properties of the ground or the sea or the atmosphere" and he noted that the proposal fitted well with his overall philosophy: "All stable processes we shall predict. All unstable processes we shall control" (cited in Fleming 2010a: 194).[4] In the period examined here, both West and East subscribed to the Promethean belief that Mastery was not only desirable but that the knowledge which underpinned it was also attainable.

Zworykin's proposal was also endorsed by oceanographer and US Army major, Athelstan Spilhaus: "In weather control, meteorology has a new goal worthy of its greatest efforts" (cited in Fleming 2010a: 194). In January 1947, with the Cold War now clearly part of the new global reality, Zworykin and Von Neumann both addressed a joint session of the American Meteorological Society (AMS) and the Institute of the Aeronautical Sciences. Measurement and calculation resulting in

---

[3] Von Neumann was a Hungarian-born, Jewish refugee, mathematician, physicist and polymath. He has become famous for his brilliance, his right-wing and militaristic views, his pivotal role in developing the H-bomb, and his foundational role in computing, amongst many other achievements. He was extremely well-connected to the military and political establishment even at the time he endorsed Zworykin's proposal.
[4] In the early nineteenth century Laplace had argued that the world was 'wholly knowable', at least in principle. Von Neumann's comments have roots in this aspect of the European Enlightenment. Similar views pervaded the Soviet bloc. As Czech dissident and later President Vaclav Havel, talking about communism, put it in 1992: communism displayed the extreme, arrogant and proud belief that "… man, as the pinnacle of everything that exists, was capable of objectively describing, explaining and controlling everything that exists" (1992).

management and control were once again themes in their presentations. Von Neumann spoke about high-speed computing in meteorology, and Zworykin spoke about weather control (Fleming 2010a: 195). Using militarily-resonant and mechanistic language, Zworykin envisaged that 'trigger mechanisms' to divert a hurricane might include covering an island with carbon black to absorb heat or generating white smoke to reflect it. 'Pity the poor Caribbean islanders', comments Fleming, 'whose tropical paradises would be invaded and possibly brutalized each hurricane season by paramilitary forces trying to save Miami' (2010a: 196).

From the start, climate intervention was of great interest to the US military. Fleming cites, as "the classic cold-war pronouncement on weather control", the commander of the Strategic Air Command, General Kenney, stating in 1947 that "[t]he nation which first learns to plot the paths of air masses accurately and learns to control the time and place of precipitation will dominate the globe" (2010a: 195–6). A telling indication of the priority given to this by the military can be seen in the fact that by 1965 the US military had 14,300 meteorologists working for it, compared to 4,500 working for the Weather Bureau (Fleming 2004: 185).

In the Soviet Union, interest in weather modification also grew after the Second World War. This was possibly in response to US initiatives and was similarly linked to a general interest in subjugating nature and opening up the Arctic zone. These efforts generated a significant scientific literature (Zikeev and Doumani 1967). In the 1950s various proposals emerged to position a sunshade over the equatorial regions to cool the planet or, more commonly, to place reflective particles in space to warm the planet, melt the permafrost and enable Siberia to flourish (Fleming 2010a: 207–8). Some of these ideas appear to have been largely speculative.

From the early 1950s Mikhail Budyko, a leading climatologist and director of the Geophysical Observatory in Leningrad (now St. Petersburg), led research programmes into the heat balance of the Earth and surface-atmosphere interactions (Fletcher 1968). There were also significant research programmes at the Institute of Geography and the Institute of Applied Geophysics, both in Moscow (Fletcher 1968: 16). Their research output is not easily accessible and some of it may have been classified. We know of at least one conference on climate modification which was held in Leningrad in April 1961 and involved all these institutes and other leading figures in meteorology (Fletcher 1968: 17).

In the USA, in 1953, a President's Advisory Committee on Weather Control was established, and in 1958 the US Congress passed Public Law 85-510 and gave the National Science Foundation (NSF) the responsibility to "initiate and support a program of study, research and evaluation in the field of weather modification" (Byers 1974: 35). Of the NSF initiatives, Harper notes that "[a]lthough such projects were couched in terms of research, their ultimate purpose was operational weather control" (2008: 22–3).

In 1958, US Senate majority leader and later President, Lyndon B. Johnson, gave a major speech after the Soviet launch of *Sputnik* and shortly before the launch of the first American satellite. Johnson is worth quoting at some length:

> Control of space means control of the world ... From space, the masters of infinity would have the power to control the earth's weather, to cause drought and flood, to change the tides and raise the levels of the sea, to divert the Gulf Stream and change temperate climates to frigid ... If, out in space, there is the ultimate position—from which total control of the earth may be exercised—then our national goal and the goal of all free men must be to win and hold that position. (cited in Fleming 2010a: 209)

The idea of climate control was overwhelmingly understood as a space of contestation between the two superpowers. The Program of the Communist Party of the Soviet Union at its 22nd Congress in 1961 included the following:

> The progress of science and technology under the conditions of the Socialist system of economy is making it possible to most effectively utilize the wealth and forces of nature for the interests of the people, make available new forms of energy and create new materials, develop methods for the modification of climatic conditions and master space. (cited in Rusin and Flit 1962: 3)

In the above two quotations we see the dominant imaginary of the times: mastery over nature is assumed to be possible, desirable and necessary; with science and technology at the heart of the drive for mastery and control. And the dominant global ideologies of each superpower, identified by Westad (2007) in their crystallised forms—liberty and justice respectively—are also visible. For the US such mastery is "the goal of all free men". For the Soviets it is "the Socialist system" and "the interests of the people".

Climate control was occasionally also seen as a terrain of global co-operation in that only co-operative efforts would generate the weather surveillance and data needed to advance understanding of the atmosphere. The World Meteorological Organisation (WMO) was formed in 1950 and became a specialised agency of the United Nations shortly thereafter. In September 1961 US President John F. Kennedy, in a speech to the United Nations, proposed "further cooperative efforts between all nations in weather prediction and eventually in weather control" (Kennedy 1961). This was only months after the Soviets had put the first man into space. That same year, Kennedy's nemesis and Stalin's successor, Soviet Premier Khrushchev, raised the issue of weather control in a report to the Supreme Soviet (Fleming 2010a: 213). A United Nations resolution the following year called for "greater knowledge of basic physical forces affecting climate and the possibility of large-scale weather modification" (cited in Fleming 2010a: 214).

An important 1962 lecture by Harry Wexler, 'On the Possibilities of Climate Control', noted that Kennedy and Khrushchev had made it 'respectable' to talk about climate modification. Wexler was the leading meteorologist in the US and chief of scientific services at the US Weather Bureau.[5] He played a key role in advancing the use of climate modelling and the deployment of meteorological satellites, and much of his work explored some of the consequences of trying to adjust the Earth's heat budget. Wexler concluded his lecture by listing, albeit not necessarily endorsing, four potential techniques for re-ordering the atmosphere. These were: to increase global temperature by 1.7°C by detonating ten H-bombs in the Arctic; to cool global temperature by 1.2°C by putting dust particles into equatorial orbit (essentially what we would today call 'solar geoengineering'); to warm the lower atmosphere and cool the

---

[5] For information on Wexler I rely heavily on Fleming (2010a, b).

stratosphere by injecting substances into space; and to destroy stratospheric ozone and cool the stratosphere by up to 80°C (Fleming 2010a: 218). Noteworthy is that the proposals had no common climate objective. Some aimed at cooling, others at warming. All aimed at control.

Note too, in all the instances cited above, the shared language and vision of Mastery held by the political leaders, the military and the scientists. The focus on climate modification emerged from the hubristic technoscientific imaginary of the times, but clearly the scientists helped to generate and sustain that imaginary. And military patronage in physics and related fields, as Forman has noted of this period, fostered a science focused on technical mastery and obsessed with gadgeteering (1987). In short, the scientific side of the many projects was inseparable from their social, political and ideological context. Research did not take place in a vacuum, and the research focus and the imaginary itself are best thought of as co-created.

Tempering Hubris

With hindsight we can also see signs of emerging doubts and of a developing awareness of a climate problem. Wexler's lecture noted that inadvertent modifications were already happening and that rising CO_2 emissions might have serious effects on the heat balance. And he was alert to growing public anxiety that, as he put it, "Man, in applying his growing energies and facilities against the power of the winds and storms, may do so with more enthusiasm than knowledge and so cause more harm than good" (cited in Fleming 2010a: 218). Indeed, by 1963 US President Kennedy had issued a secret memorandum directing that large-scale technological and scientific experiments be reviewed in advance for potentially adverse environmental effects (Harper and Doel 2010: 120–1).

Similar awareness was cautiously emerging at this time in the Soviet Union. At the 1961 Leningrad conference on climate modification already mentioned (Fletcher 1968: 17), Budyko warned that "the waste heat produced by human energy generation could, in two hundred years, rival that of the Earth's radiation balance, rendering life on Earth 'impossible'" (cited in Fleming 2010a: 236). To address the issue another scientist, M. Ye. Shvets, proposed to inject 36 million tons of particles into the stratosphere to reduce solar radiation by 2 to 3 degrees C— essentially what we would now call solar geoengineering. This was contrary to the still dominant 'unfreeze the Arctic' thinking in the Soviet Union with its underlying aim of warming.

Active work on weather modification continued throughout the 1960s and early 1970s and generated a variety of published reports. A pitch for the importance of meteorology, and the need to keep funding it, can be detected in comments by the chair of meteorology at MIT frequently cited in these reports: "Basic research in meteorology can be justified solely on the economic importance of improved weather forecasting but the possibility of weather control makes it [the research] mandatory" (cited in Keith 2000: 252). The US National Science Foundation-funded programme on 'Weather Modification' (Fletcher 1968; Byers 1974) resulted in a number of studies, papers and reports. These, increasingly, drew attention to the distant possibility of what was called 'inadvertent climate modification', which today would be called 'anthropogenic climate change', associated with rising CO_2 emissions.

Some of these ideas were considered at the highest policy level. In 1965, in one of the earliest official considerations of global warming, US President Johnson's Science Advisory Committee issued a report entitled *Restoring the Quality of our Environment*. The rise in CO_2 "… could be deleterious from the point of view of human beings", it stated in a corner of the report (in Appendix Y4), and urged the need to explore ways to restore the radiation balance, by raising the Earth's albedo. Examples cited included "… injection of condensation of freezing nuclei …" to cause cirrus cloud formation at high altitudes; and spreading small reflective particles on the surface of the oceans, at a "not excessive" cost ('rough estimates' of about $500m per year), a technique perhaps useful also in inhibiting hurricanes (White House 1965: 127). Intriguingly, therefore, one of the earliest official considerations of potential global warming recommended *only* what we would today call geoengineering! The idea of mitigation through reducing emissions was absent.

A report by J.O. Fletcher entitled 'Changing Climate' (1968) is a revealing account of both the continuing hubris (and military linkages) within scientific meteorology and atmospheric science, but tempered by an emerging awareness of both uncertainty and the possibility of climate change. The report was produced for the US military-linked RAND Corporation as part of a larger review of weather and climate control. For Fletcher, global climatic variations "… seem to be associated with variations in the vigor of the whole global atmospheric circulation, but why the global system varies is still a mystery" (1968: 6). He tentatively notes that "man is probably inadvertently influencing global climate" (p. 12) and that "some time during the next century heat pollution, like CO_2, will indeed become important on a global scale" (p. 11). This message was, in short, beginning to be made, if not heard, in the corridors of power in the US—the government, the military, and even the oil industry.[6]

Assessing research largely conducted by Soviet scientists, Fletcher concludes that interventions in particular cyclone and similar events "involve actions that we do not know how to produce efficiently". "On the other hand, various ways of influencing thermal losses and inputs to the atmosphere [the heat balance] … are much more achievable with present technology" (1968: 14). In short, and in today's terminology and using a distinction not then commonplace, climate intervention will likely be more effective than weather intervention. Fletcher states that "… to avoid undesired planetary warming, ways must be found to drain additional heat to space" (pp. 15–16). And he examines, for example, ways to create high cloud over the Arctic to increase reflectivity: "… sixty C-5 aircraft, operating from Eielson AFB [Air Force Base] and Thule AFB could deliver 1kg per km2 per day [of seeding materials] over the

---

[6] See, for example, a 1968 Stanford Resource Institute report for the American Petroleum Institute (Robinson and Robbins 1968). Drawing on the work of the President's Science Advisory Committee, it noted that exploiting all fossil fuel then thought to be recoverable would lead to concentrations of 830 ppm. The report warned that rising CO_2 would result in increases in temperature at the earth's surface, which could lead to rising seas and warming. It has been suggested the oil industry suppressed this report (see Milman 2016) and it would certainly have been in its interests to do so. But a more charitable explanation can perhaps be found in the uncertainty in the report itself, still common at that time. The report concludes by arguing that "there seems to be no doubt that the potential damage to our environment could be severe". But it then goes on to say: "Whether one chooses the CO_2 warming theory as described in detail by Revelle and others or the newer cooling prospect [from smog and industrial pollutants] indicated by McCormick and Ludwig, the prospect for the future must be of serious concern" (Robinson and Robbins 1968: 110).

entire Arctic basin" (p. 14). Fletcher concludes his report by drawing attention to the lack of understanding of climate systems:

> With environmental problems, it is convenient to think of progress in four stages—observation, understanding, prediction, and control ... Much progress is needed in all four areas in order to achieve the degree of control over climatic processes that is becoming necessary (p. 20). (Fletcher 1968)

Similar developments are apparent on the Soviet side. A remarkably prescient account for its time can be found in a paper presented by Y.K. Fedorov in the early 1970s, probably 1971, but translated and published in an edited collection in 1974, and entitled 'Modification of Meteorological Processes'. At the time, Fedorov was Director of the Hydrometeorological Service in Moscow. After recounting efforts at weather, rain and hail modification he turns to climate modification: "… modifying processes on a significantly greater scale—regional, and perhaps even global" (Fedorov 1974: 398). He stresses the complexity of the processes and gaps in knowledge, and casts doubt on earlier engineering proposals (such as the proposal to dam the Bering Strait) as "not, unfortunately, scientifically well-founded" (p. 399). He then moves onto climate change: "[i]n the not too distant future, therefore, we shall be faced with the problem, if not of global change, rather, of stabilizing the present climate, to which the economies of all the nations of the earth are geared," including compensating for "inadvertent modification" or "channel it in the direction we desire" (p. 400). Alongside limiting emissions he suggests geoengineering:

> … Theoretically, limiting the additional heat generated by human activity is not the sole means of avoiding undesirable changes in the heat budget of the planet; it is also possible to increase the other parts of the budget, such as radiation or reflection (p. 400). (Fedorov 1974)

And his focus includes albedo modification and intervention "… in processes as they evolve in the stratosphere … substantial artificial changes in the properties of the upper layers of the atmosphere, changes so great as to change the nature of the lower layers". He notes that this has been repeatedly considered by Soviet scientists (pp. 399–400). He may have been thinking of Budyko who, in his 1971 book *Climate and Life* (1974 [1971], according to Alan Robock and Philip Rasch, "proposed that if global warming ever became a serious threat, society could counter it with airplane flights in the stratosphere burning sulphur to make aerosols (small particles), similar to those found after a volcanic eruption" (IPCC 2012: 20).

By the early 1970s the US was spending tens of millions of dollars annually on weather modification research and development (Hess 1974: v). Much of the research conducted was substantially oriented to military application. The military had tested such technologies extensively throughout the 1960s as part of Projects Stormfury and Popeye, and Project Gromet in India (Harper and Doel 2010). From 1967 to 1972 the US military engaged clandestinely in extensive weather modification activities, using cloud-seeding, over the jungles of Laos and Vietnam. As part of US involvement in the wars in South-East Asia, the aim was to dramatically increase rainfall in certain areas and thereby disrupt the supply lines of the nationalist/communist guerrilla enemy. Leading US meteorologists would undoubtedly have known about such activities. They only became public knowledge in 1971–72 after details were leaked to the media (see Fleming 2010a: 177–82), and this exposure had significant consequences, as we shall see.

Retreating from the Era of Mastery

By the early 1970s some shifts in thinking were becoming apparent. The 1971 NAS-NRC report "The Atmospheric Sciences and Man's Needs: Priorities for the Future" reflects some of the changes in thinking among US meteorologists and atmospheric scientists. Significantly, this was a year after a significantly strengthened Clean Air Act was brought into force. The report notes that CO_2 increases might lead to an increase in surface temperatures of 0.5 C by 2000. "This is comparable to the natural variability of climate over 30 years; consequently, the CO_2 problem is not likely to be critical in the next few decades" (NAS-NRC 1971: 46). In addressing the 'Man's Needs' of the Report's title it retains an attachment to weather control but adds 'atmospheric quality' as a human need. The implication, and the shift in register from earlier reports, becomes clearer further on in the report as the following quotations suggest:

> In the past decade we have had to recognise that the environment can be seriously and dangerously impaired by steadily increasing industrial production and individual consumption ... Young people, especially, are scornful of a society that seems to place affluence above environmental quality. (1971: 13)
> Environmental problems will not be solved and disposed of; they will have to be managed by men and institutions that combine a wide range of technical skills with an understanding of society's needs and desires (p. 54).
> The public-policy issues that have been outlined here have much in common with similar issues arising out of such problems as management of the use of the electromagnetic spectrum or nuclear power generation or the use and control of pesticides. Problems like these, in which the technological and sociological aspects are intimately combined, are becoming more numerous and more urgent (p. 56).

Intriguingly, the report's recommendations include the need to "establish mechanisms for the rational examination of deliberate and inadvertent means for modifying weather and climate" (p. 78), and for the US to put forward a resolution to the UN "dedicating all weather-modification efforts to peaceful purposes..." (p. 79). It is hard not to conclude that the report's authors knew about, and were uncomfortable with, the still secret weather warfare being conducted in South-East Asia.

When these interventions were exposed they were widely condemned. In the US, concludes Fleming:

> The dominant opinion was that seeding clouds—like using Agent Orange or the Rome Plow, setting fire to the jungles or bombing the irrigation dikes over North Vietnam—was but one of many sordid techniques involving war on the environment that the military used in Vietnam. (2010a: 182)

The public exposure of the United States' weather warfare, combined with pressure from the Soviet Union, resulted in the United Nations adopting, in 1977, a Convention on the Prohibition of Military or Any Other Hostile Use of Environmental Modification Convention

(ENMOD). The ENMOD story is revealing of the times. Fleming provides the most detailed account and is relied upon here (2010a: 183–6). The US government under President Nixon, and the US military, were hostile to any constraints on weather modification activities. But amidst growing hostility within the US and internationally to the Vietnam War, Nixon was under pressure from a combination of leaks of information and a 1973 Senate resolution, supported by leading meteorologists, to present a resolution against hostile weather modification to the UN. In September 1974, with Nixon by now embroiled in the Watergate scandal, the Soviet Union seized the opportunity and proposed a far-reaching draft convention at the UN to forbid the use of

> meteorological, geophysical or any other scientific or technological means of influencing the environment, including weather and climate, for military and other purposes incompatible with the maintenance of international security, human well-being and health, and furthermore, never under any circumstances to resort to such means of influencing the environment and climate or to carry out preparation for their use. (cited in Fleming 2010a: 184)

This would have outlawed SGE. However, the US watered down the Soviet proposal. The final ENMOD that was adopted prohibited intentionally hostile weather and climate modification, but only at scale, and expressly permitted its peaceful use. It is generally understood that ENMOD is not a legal obstacle to pursuing solar geoengineering today.

By the mid-1970s, three decades of enthusiasm for climate modification, geoengineering and solar geoengineering were coming to an end. For the next three decades, as we shall see, the idea largely disappeared off the public policy agenda and very little scientific research into it was funded or undertaken. The instinct towards Mastery may have persisted, and indeed many of the individuals and institutions continued to operate in the environmental and climate space. But the explicit embrace of the dream of Mastery was becoming scientifically questionable and socially and politically unacceptable. The vision of Mastery went into retreat. It is to the link between this and the rising environmental movement that we now turn.

Geoengineering's Hiatus

The 1970s and Challenges to the Mastery Narrative

Why did interest in climate modification evaporate so rapidly and completely in the early 1970s? Certainly there was widespread global condemnation of US climate modification activities, and the use of environmental warfare more generally as part of the Vietnam war. The ENMOD was simply the culmination of that outrage. The Soviet Union was diplomatically astute in riding the wave of this outrage. It was also opportunistic: with lower computing capabilities and fewer resources than the US, and with less need for it militarily (its allies in the Third World mostly adopted guerrilla strategies), the climate modification race was more useful for the Soviets to constrain than to pursue. For the US, ENMOD was understood to be a restraint, hence its efforts to water down the proposed convention. But ENMOD was the aftermath, not the prime cause of geoengineering's evaporation.

Perhaps more significant was that from the early 1970s there was growing awareness within the scientific community of the complexity and inter-relatedness of the various components of the Earth's natural systems, and hence the practical difficulties of climate modification. There was also increasing knowledge about the extent and nature of the emerging climate problem. I will return, later in this Chapter, to an examination of the importance of changing thinking about, and framing of, climate change.

But the primary explanation for the end of the first wave of geoengineering can be found in a number of major shifts in the political, cultural and economic landscape. In the early 1970s the global context changed so significantly, and so rapidly, that the idea of geoengineering, rooted in an imaginary of Mastery, could no longer be sustained. We can point to the confluence of three broad developments. Firstly, the economic and fiscal crisis of the early '70s and the end of a period of plenty in the West and much of the world. Secondly, a range of crises in the 'Third World' (as it was then known) and a shift in the locus of competition between West and East, from direct confrontation to indirect, and an increasing focus by the major powers on gaining and retaining the loyalty of nations in the global South. And thirdly, a growing environmental awareness in the United States, the Soviet Union and much of the world.

Wallerstein has noted that "[w]hat had seemed in the 1960s to be the successful navigation of Third World decolonization by the United States—minimizing disruption and maximizing the smooth transfer of power to regimes that were developmentalist but scarcely revolutionary—gave way to disintegrating order, simmering discontents, and unchannelled radical sentiments" (Wallerstein 2002: n.p.). This was the period of military coups in most of Latin America and Africa, left-wing and anti-colonial wars in parts of central America, Southern Africa and south-east Asia, and Soviet efforts to contest US dominance and build its own zones of influence. This hardly cast a positive light on the 'development' and 'progress' aspects of the Cold War imaginary. Neither did the economic situation where the oil shock, petrol shortages, and sharp economic slowdown in the West heralded the end of the so-called 'Golden Age', and prompted calls to end the reliance on oil and the need for alternative, more humble, technologies.

An emergent environmental awareness also accelerated during this time growing from popular accounts of the 1960s such as Rachel Carson's *Silent Spring* (1962). 1970 saw the first 'Earth Day'. As Gilman has put it, "a radicalized environmentalist movement emerged that envisioned nature as an end in itself rather than as a standing reserve of resources to be tamed and harnessed to human ends" (2003: 245). Ironically, given they were products of 'big' science and technology, the images of this time of Earth suspended in space were widely understood to suggest a need for greater stewardship and humility. As Poole has put it: "[a]n 'eco-renaissance' took place during the Apollo years of 1968–72, framed almost exactly by the iconic 'Earthrise' and 'Blue marble' photographs four years apart" (2008: 13).

This was a period when the so-called 'proxy' Cold War was emerging, with East-West contestation increasingly played out in the Third World. Debt and conditionality became increasingly evident as mechanisms of governance, structuring relations with the Third World. There was a battle for the soul of the states of the Third World. This was the period, too, when economic instability, Cold War concerns and environmental concerns became intertwined. Third World countries were challenging the West in particular and pushing, ultimately unsuccessfully, for a New International Economic Order. Desertification and droughts in the Sahel were held responsible for creating an environment leading to the overthrow of the Ethiopian Emperor in 1974 and his replacement by a pro-Soviet, purportedly Marxist, military regime. In April 1974, in a speech to the United Nations, US Secretary of State Henry Kissinger, pushed for increased research to counter the newly-voiced threat of climate change (Behringer 2010 [2007]: 188).

In short, key elements of the Cold War imaginary—progress, development, science and technology (at least in its 'big' and militarised form), and the embrace of mastery over nature—lost much of their purchase, their "taken-for-grantedness". Gilman, borrowing from Raymond Williams, has written of "a shift in 'the structure of feeling' in the United States" (2003: 244). This was no longer a context in which deliberate climate modification could thrive, be funded, or even be respectable.

In the scientific community an early sign of this mood-shift can be found in the 1971 MIT-led *Study of Man's Impact on Climate*. This was largely a study of 'inadvertent' weather modification, the term then used to describe impacts flowing from human economic activities. The shift is immediately evident in the dedication at the front of the book. It is taken from a Sanskrit prayer: "Oh, Mother earth, ocean-girdled and mountain-breasted, pardon me for trampling on you" (SMIC 1971: frontispiece)! The study noted the increasing demands being placed on "fragile biological systems". It asked "how much can we push against the balance of nature before it is seriously upset?" (1971: 26). By the late 1970s federal funding of climate modification research and development effectively ends (Fleming 2010a: 185). But at the same time, funding is on the increase for the study of climate change.

It is unclear whether the same funding shift happened in the Soviet Union. But certainly comparable intellectual shifts are in evidence both in dissident and more 'official' circles. As early as 1968, nuclear physicist Sakharov was warning of the ecological stress being placed on the Earth by both economic systems and climate change: rising carbon dioxide levels "from the burning of coal [are] altering the heat-reflecting qualities of the atmosphere. Sooner, or later, this will reach a dangerous level" (cited in Corry 2014: 315).

Or take a 1974 publication, *Climate Changes*, by leading Russian climatologist Mikhail Budyko, which includes a speculative proposal for solar geoengineering (through injecting particles in the stratosphere). This includes estimates of the amounts of sulphur dioxide needed (15 tons), an amount 'not at all important in environmental pollution', and the suggestion that the technique would be easy and cheap. He also states:

> If we agree that it is theoretically possible to produce a noticeable change in the global climate by using a comparatively simple and economical method, it becomes incumbent on us to develop a plan for climate modification that will maintain existing climatic conditions, in spite of the tendency toward a temperature increase due to man's economic activity. (1977 [1974]: 244)

Whilst proposing intervention, and assuming perfect 'knowability', Budyko also goes on to add a note of caution:

> The perfection of theories of climate is of great importance for solving these problems, since current simplified theories are inadequate to determine all the possible changes in weather conditions in different regions of the globe that may result from modifications of the aerosol layer of the stratosphere. (Budyko 1977 [1974], cited in Schneider 1996: 293)

Budyko remained a proponent of solar geoengineering for the rest of his life. But here he shifts the argument for climate modification from one of desirability as an indicator of human mastery, to one of necessity. And, whilst retaining some of the older vision, climate modification is couched not in terms of 'improving' the climate, as the Mastery discourse would have done, but in terms of 'maintaining' the existing climate.

A comparable article on the topic, although taking the cautious tone of Budyko a step further, can be found in an article in *Science* by leading US climatologists William Kellogg and Stephen Schneider entitled 'Climate Stabilization: For Better or for Worse?' (1974). They note that climate prediction will not be straightforward given that "… our climatic system is a highly nonlinear, interactive system that has defied a complete quantitative description" (p. 1163). They argue that even if any future climate could be predicted, "control would be a hazardous venture" (p. 1163). In a mark of the times their article also refers to contemporary debates around the Club of Rome's influential *Limits to Growth* publication (Meadows et al. 1972), and other "advocates of reduced economic growth" (p. 1167). Given that "mankind will very likely have the technological capability to alter climate purposefully as well as inadvertently" (p. 1168), they analyse a range of climate modification schemes, including particle injections to counteract warming, to "create a kind of 'stratospheric smog'" (p. 1170). Their prescient account, prefiguring arguments which emerge in the current wave of interest in geoengineering, leads them to be sceptical of climate modification proposals on three grounds: because of an inability to "adequately foresee the outcome"; because it would lead to conflict since "cause and effect are hard to unravel"; and because of uncertainty as to how the decision would be made regarding "whose climate should be preserved, whose improved, and whose sacrificed" (pp. 1170–1).

For the remainder of the 1970s and through until the end of the Cold War around 1990, very little other work on geoengineering is published. There is a speculative article by Marchetti entitled 'On geoengineering and the CO_2 problem' (1977) published in the first issue of a new journal, *Climatic Change*. He argues not for solar geoengineering but for the need to consider capturing and injecting CO_2 into the ocean depths, an early version of carbon capture and storage (CCS). Dyson and Marland (1979) use 'old-school' language in entitling their paper 'Technical fixes for the climatic effects of CO_2'. But the paper itself focuses on reforestation. Broecker (1985) touches on ocean fertilisation to draw down CO_2. Budyko in the Soviet Union is almost alone in remaining focused on solar geoengineering: in *The Earth's Climate: Past and Future* (1982) he calculates the tonnage of sulphuric acid needed to reduce solar radiation by 1%.

From the early to mid-1970s until the middle of the first decade of the 2000s geoengineering basically disappeared from the policy agenda. One can think of this three decade-long period as a Hiatus, dividing the First Wave of interest in geoengineering from the Second Wave of interest when it re-emerged around 2005. This Hiatus comprised two distinct stages. First, from the early 1970s to the end of the Cold War, discussed above, when geoengineering was essentially not thought about and was Unimagined or largely disappeared from the imaginative horizon. Then, during the 1990s until the early 2000s, when some attempts to revive the idea of geoengineering were unsuccessful, what had been unimagined became, in effect, 'taboo'.

Attempts at Post-Cold War Revival

The Cold War came to an end around 1989–90 with the implosion of the Soviet Union and many of its client states. This immediately meant that countries of the global South could no longer play the two great powers off against one another, or rely on Soviet support. It also meant that capitalism was seen as economically triumphant. The US ideology of 'liberty' had trumped the Soviet one of 'justice'.

In the new, market-friendly, world one might have expected a revival of interest in geoengineering. After all, solar geoengineering, long held to be the cheapest option for tackling global warming, should have been attractive. And there were indeed some attempts along these lines—emerging, perhaps predictably, from Edward Teller and his colleagues at the Lawrence Livermore National Laboratory (LLNL).

The most notable effort to revive the idea emanated from parts of the US scientific and policy community. A major 1992 National Academy of Sciences (NAS) study on *The Policy Implications of Greenhouse Warming: mitigation, adaptation and the science base* included a chapter on geoengineering. Largely written by Robert Frosch and the Livermore group of scientists, this Chapter assumed that geoengineering could mitigate greenhouse gas effects and do so affordably, and focused on whether this could be done with acceptable or manageable adverse effects (NAS 1992: 434). With regard to solar geoengineering, which it called "stratospheric dust", the report notes that the chemical reactions might destroy ozone: "a possible side effect that must be considered and understood before this possible mitigation option can be considered for use" (p. 451).

This appears to be the first study which places so much emphasis on solar geoengineering's purported cheapness. Apart from this, the bulk of the chapter's examination of solar aerosol geoengineering is devoted to the comparative costs of distributing the 'dust' by navy guns, rockets, or balloons. There is even consideration of emissions being done by dedicated ships roaming the oceans, or by purpose-built, high-sulphur coal burning power stations placed at sea! A 1992 cartoon from a popular publication, *Geographical Magazine*, and titled 'Shooting dust into the stratosphere to offset global warming' suggests the idea was lampooned at the time (reproduced in Fleming 2006: 238). Apart from this however, the NAS report's chapter on geoengineering generated little interest and the report's overall focus, as its sub-title suggests, was elsewhere. The report summary recommended research into geoengineering options but made explicit that "[t]his is not a recommendation that geoengineering options be undertaken at this time" (1992: 81).

Leading climate scientist Stephen Schneider, who was part of the panel that compiled the 1992 study has recalled that "… the very idea of including a chapter on geoengineering led to serious internal and external debates", with critics (himself included) arguing it would be used as "an excuse to continue polluting". Another panellist and the report's lead author, Robert Frosch, apparently argued that if climate change was more extreme than expected "[w]e would simply have to practise geoengineering as the 'least evil'". Schneider seems to have been persuaded by this argument as he continues:

> Although sceptical about the viability of specific engineering proposals and the questionable symbolism of suggesting that we could sidestep real reductions in emissions, I nonetheless voted reluctantly with the majority of the NAS panelists who agreed to allow a carefully worded chapter on the geoengineering options to remain in the report. (Schneider 2001: 418)

The arguments we will meet in the mid-2000s with geoengineering's re-emergence are all prefigured here. It is also apparent that the idea was still effectively taboo. A thought bubble was being floated to see if it would fly. It did not.

Little other scientific work was being done in the area, with Charlson et al. (1992) a rare example. Russian climatologists, Mikhail Budyko and Yuri Izrael, also published new work on climate and solar geoengineering (1992 [1987]). Long-time enthusiast for solar geoengineering, Lawrence Livermore scientist Michael MacCracken, in his 1991 paper to a limited attendance workshop, acknowledges that geoengineering and its solar variant, which he likens to "a human volcano", is "a particularly controversial option" (1991: 2). A 1992 article, 'Taking Geoengineering Seriously', by Keith and Dowlatabadi was largely an argument, "ignoring important ethical issues", for research money to be allocated to geoengineering: essential "unless nasty surprises are assigned a zero probability" (1992: 293). Their article is an early example of what would become a common trope: framing solar geoengineering as humanity's 'Plan B'.

There is remarkably little published research beyond this. The early 1990s effort to revive interest is largely ignored and, in effect, geoengineering remains beyond the realm of respectable policy consideration. The first assessment report of the IPCC was published in 1990 and made no mention of geoengineering. Neither did the United Nations Framework Convention on Climate Change (UNFCCC) adopted at the Earth Summit in 1992, nor the subsequent Kyoto protocol.

In the mid-1990s geoengineering briefly re-emerged in debates between scientists, as concern grew about the warming effects of increasing GHG concentrations and the apparent lack of action to tackle this. But these debates, whilst not secret, were not influential on climate policy debates more generally. In July 1996, Issue 3 of the journal *Climatic Change* was devoted to the topic. It drew on papers presented to a 1994 symposium by scientists and non-scientists including some traditionally hostile towards geoengineering and committed to the then-dominant climate strategy of mitigation through abatement. The titles are revealing of the tentativeness with which the topic was approached.

The editorial, 'Could we/should we engineer the Earth's climate?' stressed that the various articles "... provide analysis not advocacy ... [and] make it abundantly clear that relying on these approaches now would be irresponsible, given our current state of understanding of the climate system" (Marland 1996: 277). Stephen Schneider's article 'Geoengineering: could—or should—we do it?' concluded that:

[A]lthough I believe it would be irresponsible to implement any largescale geoengineering scheme until scientific, legal and management uncertainties are substantially narrowed, I do agree that, given the potential for large inadvertent climatic changes now being built into the earth system, more systematic study of the potential for geoengineering is probably needed. (1996: 291)

The more enthusiastic Thomas Schelling article talks about geoengineering being "an unmentionable subject" and the need to pull it "out of the closet, but I'll wager not all the way" (1996: 303). A legal academic, Bodansky, in his article entitled 'May we engineer the climate?' notes that "[w]hether we can engineer the climate will depend not only on its technical feasibility, but also on its political acceptability" (1996: 320).

In the wake of the *Climatic Change* articles a number of scientific enthusiasts published work favourable towards solar geoengineering. These included scientists from the military-scientific Livermore Laboratory (LLNL), including an ageing Edward Teller, his associates Lowell Wood, Michael MacCracken and others (Teller et al. 1997; Teller 1997; see also Flannery et al. 1997). Teller et al.'s paper argued that solar geoengineering would be much cheaper than mitigation and was technically possible. Indeed, the paper frames the issue in terms comparable to Teller's over-confidence of old: "Today, our scientific knowledge and our technological capability already are likely sufficient to provide solutions to these problems; both knowledge and capability in time-to-come will certainly be greater" (Teller et al. 1997: 17).

But, as Schelling had predicted, this was not enough to get the taboo lifted from policy discussions of geoengineering, to get it 'out of the closet'. Indeed, enthusiasts like these may have helped keep it in. The IPCC's Second Assessment Report, Working Group 2, mentions solar geoengineering in passing within a brief section on counter-balancing climate change, in large part authored by Michael MacCracken (1995b: 812). In the synthesis summary of the whole second assessment report, the formulation is dismissive: geoengineering is "…likely to be ineffective, expensive to sustain and/or to have serious environmental and other effects that are in many cases poorly understood" (IPCC 1995a: 41). The IPCC's Third Assessment Report in 2000 is also cursory and equally dismissive when it argued that "most papers on geo-engineering contain expressions of concern about unexpected environmental impacts, our lack of complete understanding of the systems involved, and concerns with the legal and ethical implications … Unlike other strategies, geo-engineering addresses the symptoms rather than the causes of climate change" (IPCC 2000: 334).

Political unacceptability was apparent when, in September 2001, President George W. Bush's Climate Change Technology Committee included geoengineering in its initial formulations of possible approaches. A high-level meeting of about a dozen scientists was held with some, such as Lowell Wood, on speakerphone. According to David Keith, "[i]t was a frank discussion of geoengineering options and the need for research funding" (reported by Goodell 2006: n.p.). The idea was dropped by President Bush's team, according to another participant, as it might add "… to the perception that the Administration's emission reduction program was not serious" (MacCracken 2006: 239). In any event, the September 11th attacks on the US only days later pushed climate and many other issues off the US political agenda.

Why Attempts to Revive SGE Failed

In many respects it seems surprising that geoengineering did not re-emerge as a serious policy option in the aftermath of the Cold War. After all, the end of the Cold War is closely linked to the acceleration of 'globalisation' and an expansion of the neoliberal project from its US heartland. Central tenets of neoliberalism included the primacy of economic growth and the removal of trade barriers, individual consumer choice, reduced government regulation of business practices, and a development model entailing 'The Rest' following 'the West'.

'Globalisation' became a framing metaphor not only for the neoliberal economic project, but also across a range of disciplines and discourses, including climate and environmental discourses. Accompanying these developments was an increase in the perceived and actual power of global corporations. There was a growth too of what Milonakis and Fine (2009) call 'economics imperialism': an elevation of the status of economics as a discipline, and of the perceived salience of economic metrics in making policy choices, and "the extension of economic analysis to subject matter beyond its traditional borders" (2009: 5). In practice a particular notion of economics, rooted in neo-classical economics, centred on so-called 'rationalism' and often manifesting as cost-benefit analysis, assumed centrality in policymaking, even being adopted as a worldview and method by many in the social sciences.

One might therefore have expected geoengineering to be embraced following the end of the Cold War, and solar geoengineering in particular with its suggestion of being a cheap 'solution' to climate change. The 1992 NAS study did indeed put much emphasis on solar geoengineering's purported cheapness. The report does not make explicit solar geoengineering's compatibility with the neoliberal project as a whole which was then in a triumphalist phase in the immediate aftermath of the Cold War. But this was clearly part of the thinking, as an interview at the time with Robert Frosch reveals. Frosch, previously with NASA and also a vice-president at General Motors, led the compilation of the NAS report's chapter on geoengineering. Interviewed by *Newsday* he said:

> I don't know why anybody should feel obligated to reduce carbon dioxide if there are better ways to do it. When you start making deep cuts, you're talking about spending some real money and changing the entire economy. I don't understand why we're so casual about tinkering with the whole way people live on Earth, but not tinkering a little further with the way we influence the environment. ... I think of it as designer volcanic dust put up with Jules Verne methods. (cited in Fagin 1992: 7)

The failure of the attempts through the 1990s, and again in 2001, to resurrect geoengineering can be ascribed to a number of factors, both small and large: the credibility of the people proposing the idea; the playing down of climate change in the corridors of US power, alongside the increasingly dire predictions of the climate scientists and their commitment to mitigation; and a belief that environmental problems could be solved alongside the neoliberal project now that the 'end of history' had arrived. I will briefly touch on these.

Firstly, most of the enthusiasts for solar geoengineering during the 1990s were individuals strongly associated with weapons development and with the old Mastery paradigm (for example Edward Teller and Lowell Wood), and from institutions of 'big science' such as the Lawrence Livermore Laboratory (for example Michael MacCracken). Those with weaker links to these institutions, and indeed much of the climate science community, were less enthusiastic (for example Stephen Schneider). In short, geoengineering's scientific messengers lacked social and environmental credibility, even as they were scientifically respected. There was a growing divergence between their values and the increasing ecological awareness of many in the epistemic community of climate science. This may, in part, have been generational.

Secondly, in the early 1990s in the United States, whilst climate change was recognized as an issue, it was beginning to be a contested one. President George H.W. Bush had come into office in 1988 pledging to be the "environmental president" and to "do something about the greenhouse effect". In office he was less active than expected and his administration was accused of playing a major role in watering down both the Rio Declaration and the UNFCCC which emerged from the first Earth Summit in 1992 (Carcasson 2006). This was a manifestation of an emerging strand of thinking within the Republican side of US politics. During the early 1990s President George H.W. Bush began to see "the economic costs of an aggressive stance on global warming as too high for comfort" and his Chief of Staff was one of the earliest leadership figures to question the veracity of the climate science (Philander 2008: 146–7). In a foretaste of future thinking, junior officials in his administration notoriously edited NASA climatologist James Hansen's 1988 testimony to Congress to downplay his conclusion that global warming was accelerating. The 'merchants of doubt' were becoming increasingly active (Oreskes and Conway 2010).[7] Why consider geoengineering, or indeed any action, if the science was wrong?

---

[7] As we now know, at this time a range of think-tanks were switching from the now disappeared 'red peril' to the 'green peril' of environmentalism, and a range of corporate interests were starting to promote doubt about global warming (Oreskes and Conway 2010).

The flipside of this was that whilst the US political commitment to action was softening, warnings about the problem of global warming were hardening amongst climate scientists. Not only were the climate prospects dire but mitigation through cutting greenhouse gas emissions was seen as the obvious policy response. Adaptation, at this time, was regarded as an adjunct policy at best, necessary since actual emissions reductions were proving inadequate. Indeed, for much of the period covered in this chapter adaptation was regarded with hostility by many climate scientists, environmentalists and climate policymakers. Al Gore famously argued in 1992 that adaptation represented a "kind of laziness, an arrogant faith in our ability to react in time to save our skins", a position he later retracted (cited in Pielke et al. 2007: 597). These would prove to be formative years in shaping the political-environmental outlook of the epistemic community of climate science—making it green-tinged and, occasionally, activist.[8] In short, whilst those on the right of US politics were shifting away from action on climate, the climate scientists themselves were becoming increasingly shrill in their warnings of looming catastrophe and their calls for deep cuts in emissions. Why consider geoengineering if a policy solution already existed? Indeed, geoengineering could only result in us taking our eye off the mitigation ball.

Finally, and relatedly, there was another reason keeping geoengineering unimaginable at this time: a brief moment of naïve optimism regarding the ability of the 'global community' to address climate change. The Cold War had ended with the triumph of the West. Global conflict had come to an end with only one superpower left standing. It was believed that North-South problems could now be addressed free from ideology and the competing demands of the post-war past. Fukuyama's 'end of history' thesis held that no serious ideological alternative to liberal democracy and market capitalism now existed (1989). It was a perspective widely embraced by the advocates of neoliberal globalisation.

The elite vision of the time was that global problems, including a healthy environment, could be solved through global co-operation.

---

[8] I follow Litfin's more expanded use of Haas's 'epistemic communities' notion, and her emphasis on paying particular attention to how "discursive practices promote specific narratives about social problems" (1995: 252).

Geoengineering, with its implication of unilateral global action, was incompatible with this. Further, the 1992 Rio Earth Summit, alongside a vision of sustainable development, had introduced a framework of action against global warming (the UNFCCC), essentially committing the industrialized countries to reduce their greenhouse gas emissions using market mechanisms. The Earth Summit placed mitigation at the centre of its climate thinking, and given that it promoted the virtues of a healthy environment perhaps for the first time in so authoritative a document, many saw in 'sustainable development' a new and progressive narrative. Which it was, to a certain extent (Robinson 2004): sufficiently so to keep geoengineering unimaginable.

Bernstein and Ivanova have noted how the Earth Summit tried to capitalize on the 'new optimism' which, its organizers hoped, "… would be a catalyst for a post-cold war order characterized by an open, market-friendly international economic system and a peaceful multilateral political system" (2008: 162). But they also note that, in hindsight, the commitment to sustainability was "…less significant than its integrative proposition that action on the global environment rested on a foundation of liberal economic growth" (p. 164). Illusions about the 'end of history' and optimism about tackling the climate challenge were not to last. But until then there was no reason to treat geoengineering as anything other than Taboo.

In short, the geoengineering idea failed to take off in the aftermath of the Cold War because its proponents were not credible, because it made little sense to both the 'do-nothing' right and the 'mitigation' left in US politics, and because of a misplaced optimism (or no-alternative 'realism') about the possibility of addressing climate change using market mechanisms and global treaties.

Shifting Imaginaries of Climate, Nature and Globalisation

I have shown how we can usefully think of geoengineering's First Wave as having successive phases. Its Mastery phase was succeeded by a Hiatus in which solar geoengineering was, at first, not much thought about. When it was, in the aftermath of the Cold War, it was not given serious consideration as a policy option and even came to be regarded as Taboo. First Wave geoengineering had a practical dimension, as with the attempts to change weather and regional climates in the 1960s and 1970s. But throughout this period it was largely a speculative technology, an idea which was embraced by some and rejected by others. Of necessity it was grounded in a range of assumptions about the desirability and possibility of re-making the Earth and re-shaping the climate.

The master narratives and reigning imaginaries changed over the sixty-year period we have explored in this Chapter. Values shifted. Power relations changed. So too did knowledge and understanding of the climate system. All these changes affected how the idea of geoengineering as a practical proposal was conceived and received, and its presumed viability as a policy option. Changes, both material and ideational, in three areas can help us understand why there was a turn away from the idea of Mastery and a movement towards treating geoengineering as taboo. Two have an inflection point in the 1970s: the discovery of global warming, and the rising prominence of environmentalism. The third, associated with the 1990s, is the optimistic turn towards globalisation in the immediate aftermath of the end of the Cold War. Taken together, these also help us understand the context within which SGE was to re-emerge in the mid-2000s.

Changing Climate Science

Global warming was 'discovered' in the early 1970s (see for example Behringer 2010 [2007]; Weart 2008), roughly coincident with the end of the first wave of interest in geoengineering. Although the physical basis of the greenhouse effect was understood in the nineteenth century and data on CO_2 concentrations was available from the 1950s, the dominant view until the early 1970s, and with some support from the temperature record, was that the global climate was cooling (Behringer 2010 [2007]: 185). Prior to this, whilst there were some scientists, as I have shown, alert to the possibility of warming, this was not the agreed, nor even the dominant view. As Budyko, among the leading climatologists of the time, put it in his 1971 book *Climate and Life*: "there is at the present time no generally accepted view on the causes of climatic change and fluctuation" (1974 [1971]: 292). The cutting-edge 1971, global but MIT-led, *Report of the Study of Man's Impact on Climate* involved some of the world's leading climatologists from both East and West. It asserted that "… we do not yet know enough to make positive assertions about man's potential role in climate" (SMIC 1971: 3).[9] The Club of Rome's influential publication *Limits to Growth* (Meadows et al. 1972) mentioned CO_2 but included no concept of global warming. Only by the mid-to-late 1970s was the view becoming established amongst the scientific community that the CO_2 effect was significant, that it was causing atmospheric warming, and with it the insight that global warming posed a potential threat (Behringer 2010 [2007]: 190).

Almost concurrently, the military-industrial-scientific complex, which was at the heart of the first period of the Cold War, was facing scrutiny and criticism. Numerous studies have shown how stable and co-dependent the relationship was between the state, military strategists and the scientists and scientific institutions for much of the Cold War (see for example McNeill and Unger 2010: 14; Fleming 2010a; Oreskes and Krige 2014; Miller 2004). Meteorology and the climate sciences more generally were no exception, as is evidenced not only in their funding but also in their being at the forefront of the application of emergent satellite and computer technologies. By the early 1970s, at around the same time as global warming was being discovered, many scientists were accused of "… selling their knowledge for immoral causes and of sacrificing science's freedom for personal advantage" (McNeill and Unger 2010: 17). The charge cut deeply and some scientists, in the US and also in the Soviet Union, sought to re-establish some of the academic distance they had pre-War.[10]

---

[9] This intriguing study prefigures the idea of the Intergovernmental Panel on Climate Change (IPCC), not only in its participants, but also in its call for "…an international consensus among concerned scientists…" (SMIC 1971: 3–4). The reports instincts were to seek non-anthropogenic causes for climate change, and, if anything, to link human industrial activities and attendant pollution with recorded cooling of 0.3 degrees in the Northern hemisphere between 1940 and 1970 (p. 10).

[10] The story is in fact slightly more complex and, from the 1960s, in the United States, a political bifurcation starts to emerge. The biology-centred disciplinary side was focused on ecological prob-

In addition, by the mid- to late-1970s the 'epistemic community' (Haas 1992; Litfin 1995) of the emerging theorization and science of global warming was being driven by its findings to assume a more critical social stance. Not only was the connection between warming and anthropogenic emissions emerging, explicitly linking the 'social' with the condition of the 'natural'. But it was also not a major argumentative leap to point out the implication that something was amiss with 'our civilization'. Further, through the 1970s and into the 1980s, scientists were starting to reach uncomfortable, and increasingly alarming, conclusions about the possible local and regional effects of a continuing rise in emissions, and not only in the long term. Scientists and scientific knowledge had been conferred an elevated status in society. So it was difficult, at least at this time, to question the increasingly critical comments emerging from the climate scientists. As McNeill and Unger have put it, writing about this period: "[I]t was difficult for politicians and strategists to question those statements without risking their own credibility" (2010: 17).

Starting in the 1970s and running through until the late 1980s a number of epistemological shifts occurred within climate science. There was growing recognition of the complexity of the climate system and increasing reference to feedback loops, hysteresis, non-linearity, chaotic interactions, probabilistic assessments and so on. Whilst greatly expanding their knowledge and modelling activities, climate scientists were also aware of major gaps in their understanding. This was not humility, but it was in stark contrast to the early geoengineers' confidence in how much they knew. Understanding the climate as a complex, dynamic system rather than an easily modifiable machine made arguments for geoengineering less persuasive and affectively homeless. Geoengineering did not sit comfortably with the new 'structure of feeling' which emerged during the 1970s.

---

lems and publishing openly. The geophysics-centred side remained, in large part, more directly responsive to the military's operational needs, and its key findings were often classified (Doel 2003). It was largely the former that gravitated towards a more environmentally aware politics, and only some of the latter. As Masco argues: during the Cold War and beyond, the damaged biosphere was "[b]oth discovered as an object of state interest and repressed as a political project" (2010: 17). This may also help explain why scientists from the military-linked Lawrence Livermore Laboratory (LLNL) could never be credible promoters of solar geoengineering.

The evolution of 'climate change' as an object of science and policy helps us to understand why geoengineering largely dropped off the policy agenda from the early 1970s and the later reluctance to lift the geoengineering taboo. It helps explain the emergence, especially from the late 1980s, of an increasingly vocal 'epistemic community' of climate scientists committed to greenhouse gas abatement (mitigation) as a central strategy; and, as we shall see in subsequent Chapters, it assists our understanding both of geoengineering's current move into the mainstream of climate policy consideration, as well as the widespread scientific reluctance to embrace it.

Détente and Environmentalism

A shift in the Cold War during the 1970s, from direct to indirect competition, is also relevant to understanding the shifting imaginaries of the time. The early 1970s was the so-called *détente* period of the Cold War between the US and the Soviet Union. This saw a shift towards proxy Cold War conflicts in the 'global South' and, as indicated earlier in relation to Ethiopia and India, a concern in the West about environmental problems provoking revolution. It also meant politicians in both East and West had to search for issues to bring into dialogue if *détente* was to be given concrete form. Environmentalism was on the rise in the West and the environment was one of the few areas in the 'East' where some oppositional activity was tolerated (Corry 2014). Not surprisingly, as Hünemörder has put it: "[i]nvoking the slogan 'Only One Earth' enabled the two superpowers to bridge, at least partially, the deep ideological gap dividing them" (Hünemörder 2010: 274; see also Uekötter 2010: 344).

The policy of *détente* was occurring concurrently with an emerging environmentalism. Both opened a significant political space which made possible a normative convergence between the emerging environmental movement and the epistemic community of climate science. The former had been largely hostile to 'big science' and focused on problems in the ambient environment (such as pollution), but now found increasing validation of their larger concerns in much of the new climate and ecological science. The latter were keen to assert some relative autonomy from the military-industrial complex and increasingly uncomfortable with the implications of their own findings.[11] The net effect of the shift to a more independent stance already described, and the space opened by environmentalism and détente, was to make an attachment to science more central to the environmental movement, and to 'green' (at least to some extent) the normative sea in which the climate scientists swam. It was no longer a context conducive to Teller-style talk of asserting 'control over nature'.

The 1970s was the period when most of the shifts described above occurred. These help explain why geoengineering fell off the policy table at that time and the deep shifts in values and knowledge which ensued made it difficult for solar geoengineering's boosters to rebuild interest in their idea at the end of the Cold War. Indeed, by the 1980s geoengineering was essentially taboo.

Globalisation and Universalist Discourses

Developments in the 1990s were also important in understanding why the Taboo could not easily be lifted. The dominant master narrative (and practical project) of the 1990s was 'globalisation', understood as the expansion of neoliberalism globally. New multilateral initiatives commenced aimed at creating institutions able to take this vision forward. The World Trade Organisation (WTO) emerged in 1995. On the environmental front, as we have already touched upon, the major initiative was the 1992 Earth Summit and its attendant commitment to an "open international economic system that would lead to economic growth and sustainable development in all countries" (Principle 12) and to action on climate change. Sustainable Development was overwhelmingly attached to the global rollout of the neoliberal project, operationalised in the global South as the Washington 'consensus' (see Steger 2009: 77 for an account of this). At the time, from an environmental perspective, this was either believed to be compatible with enhanced environmental protection and climate action, or regarded as a framework for which,

---

[11] Later this would manifest in prominent establishment scientists, such as NASA's James Hansen, becoming activist in relation to climate policy.

2 Geoengineering's Past: From Mastery to Taboo 65

after the collapse of communism, 'there was no alternative'. It also meant little appetite for pursuing nationally-based projects of solar geoengineering and lifting the effective Taboo.

This period saw the emergence of a number of globalised discourses. Poverty, development and the environment were reframed as *global* problems requiring *global* solutions. Climate was no exception. Global warming was also increasingly understood as a global challenge requiring global action, and this had paradoxical implications for the acceptability of solar geoengineering.

We have seen how from the 1970s to the early 1990s there was a shift in framing the object of concern from local/regional 'weather' to global 'climate', and 'global climate change' in particular. During much of the Mastery phase and even into the 1970s, as Miller has demonstrated, "[c]limate remained merely another way of describing the weather, a statistical artifact constructed through mathematical averaging" (Miller 2004: 52). Hence, the First Wave geoengineering fantasies and proposals covered both weather control and climate control, and the boundaries between these objectives, and also the scale at which they were to be enacted, were never clear.

Moving from 'weather' to 'global climate' entailed treating the atmosphere as a single entity and, increasingly, as linked to all aspects of what we now term 'Earth Systems'. The term 'climate', as Miller puts it, went 'from signifying an aggregation of local weather patterns to signifying an ontologically unitary whole capable of being understood and managed on scales no smaller than the globe itself' (Miller 2004: 54). Climate change was to be understood as a global risk needing global science and global political co-operation. By the end of the Cold War, the imaginary this fed into and helped generate, was of the need for a *global* politics able to tackle a global problem by global action: reducing emissions in order to limit disruption of the atmosphere.

* * *

The imaginaries of geoengineering change across the Mastery, Unimagined and Taboo phases. Table 2.1 summarises, schematically and in simplified form, the master narrative of each phase, as well as the mood of the time.

Table 2.1 Schematic summary of reigning geoengineering imaginaries (1945–2005)

	Mastery (1945–1970s)	Unimagined (early 1970s–1990)	Taboo (1990–2005)
Holding imaginary	Progress Development Titanic struggle against other superpower and its vision (liberty vs. justice)	Progress Development 1.0 Environmental limits Liberty vs. Justice conflict played out in Third World	Markets Development 2.0 Environmental Risk 'There is no alternative'
	MODERNISATION THROUGH STRENGTH	MODERNISATION THROUGH LEADERSHIP	MODERNISATION THROUGH GLOBALISATION
Nature/environment	Human mastery over/subjugation of nature both possible and desirable	Heightened awareness of environmental problems/impacts/damage; Contestation: human mastery over vs. human responsibility for	(Pol'ticised) awareness of environmental problems Environment + Development = Sustainable Development possible
Science & Technology	Faith in authority of science and scientists; 'Big' S&T (including hubristic projects) Close to military-industrial complex	Faith in authority of science and scientists; Reduced faith in 'big' science	Challenges to authority and certainty of science; Increased 'speed' of technological change
Climate	No weather-climate distinction Something to be modified; No generally understood warming trajectory	Emergence of warming as a (possibly serious) problem	Globalisation of climate Anthropogenic link; Cost of action Emergent climate 'denialism'
	WEATHER? CLIMATE?	GROWING KNOWLEDGE	GROWING CONCERN
Mood	Positive, confident, unbridled optimism	Tempered optimism	Optimism then rising pessimism despite 'end of history'; Uncertainty
Geoengineering implication	A 'solution' without a problem	No imagined problem needing geoengineering	Emerging climate problem, no credible SGE advocate

It also summarises the particular dominant understandings of nature/environment, and of technology in each phase. Finally it captures the dominant scientific (and political) understanding of climate and the climate problem in each of the phases, and the implications for geoengineering. These are mainly the reigning imaginaries in the United States, which is where (other than in the Mastery phase) almost all thinking and research relevant to geoengineering was occurring.

In the Mastery phase, geoengineering was a 'solution' without an obvious problem. It was simply one manifestation of the desire to tame nature, and spurred the invention and exaggeration of 'problems' such as too much ice in the Arctic or the need to combat desertification in Africa. It was embraced because it could be done, not because it was needed.

Geoengineering's Hiatus covers two distinct phases. From the early 1970s until the late 1980s, geoengineering does not make sense as a solution to anything. With environmental limits acknowledged but the magnitude of potential warming not widely understood, and with climate policy low in the list of public policy concerns, it could be said that no problem was imagined which could throw up geoengineering as a solution.

In the aftermath of the Cold War, and with growing concern about climate change, geoengineering peeks out of the closet. But given the background of its sponsors, the primacy of as-yet-untested mitigation strategies, the shifting outlook of the epistemic community, and the difficulties of getting global climate policy in place, there is no appetite by either scientists or politicians for lifting the taboo on a controversial idea.

Optimism about what global co-operation might achieve was not to last. Despite the end of the Cold War ushering in an era of globalising, universalist discourses, it soon became apparent that the imagined solution to the global climate problem—cutting back on emissions—sat uneasily with the master narrative of the time: market globalism and the unconstrained expansion of economic growth and the neoliberal project globally.

It was into this context that calls to lift the Taboo on solar geoengineering started to be heard in the early 2000s. If emissions reductions were not occurring, if the climate prospects were worsening, and if the broader neoliberal project was to be promoted (or at least be seen as unassailable), then perhaps solar geoengineering needed another look. It was a 'solution', a way to cool the planet that could fit neatly with such assumptions. In the next Chapter we turn to the re-emergence of geoengineering, its Second Wave, commencing in about 2005. The continuities and discontinuities with these earlier phases will become apparent.

References

Behringer, W. (2010 [2007]). *A cultural history of climate*. Translated from German by P. Camiller. Cambridge: Polity.
Bernstein, S., & Ivanova, M. (2008). Institutional fragmentation and normative compromise in global environmental governance: What prospects for re-embedding? In S. Bernstein & L. W. Pauly (Eds.), *Global liberalism and political order: Toward a new grand compromise?* New York: State University of New York Press.
Bodansky, D. (1996). May we engineer the climate? *Climatic Change, 33*(3), 309–321.
Bonnheim, N. B. (2010). History of climate engineering. *WIREs Climate Change, 1*(6), 891–897.
Brain, S. (2010). The Great Stalin Plan for the transformation of nature. *Environmental History, 15*(4), 670–700.
Broecker, W. S. (1985). *How to build a habitable planet*. Palisades, NY: Eldigio.
Budyko, M. I. (1974 [1971]). *Climate and life*. Translated from Russian and edited by D. H. Miller. New York: Academic Press.
Budyko, M. I. (1977 [1974]). *Climate changes*. Translated from Russian. Washington, DC: American Geophysical Union.
Budyko, M. I. (1982). *The Earth's climate, past and future*. New York: Academic.
Budyko, M. I., & Izrael, Y. A. (1992 [1987]). *Anthropogenic climatic change*. Leningrad: Hydrometeoizdat. In Russian; English edition (1992) Tucson, AZ: University of Arizona Press.
Byers, H. R. (1974). History of weather modification. In W. N. Hess (Ed.), *Weather and climate modification* (pp. 3–44). New York: John Wiley.
Carcasson, M. (2006). Prudence, procrastination, or politics: George Bush and the Earth Summit of 1992. In M. J. Medhurst (Ed.), *The rhetorical presidency of George H. W. Bush* (pp. 119–148). College Station, TX: Texas A&M University Press.
Carson, R. (1962). *Silent Spring*. New York: Fawcett Crest.

Charlson, R. J., Schwartz, S. E., Hales, J. M., Cess, R. D., Coakley Jr., J. A., Hansen, J. E., et al. (1992). Climate forcing by anthropogenic aerosols. *Science, 255*(5043), 423–430.
Corry, O. (2014). The green legacy of 1989: Revolutions, environmentalism and the global age. *Political Studies, 62*(2), 309–325.
Doel, R. (2003). Constituting the postwar earth sciences: The military's influence on the environmental sciences in the USA after 1945. *Social Studies of Science, 33*(5), 635–666.
Dyson, F. J., & Marland, G. (1979). Technical fixes for the climatic effects of CO_2. In *Workshop on the global effects of carbon dioxide from fossil fuels*, Rep. CONF-770385 (pp. 111–118). Washington, DC: U.S. Department of Energy.
Fagin, D. (1992, April 13). Tinkering with the environment. *Newsday*.
Fedorov, Y. K. (1974). Modification of meteorological processes. In W. N. Hess (Ed.), *Weather and climate modification*. New York: Wiley-Interscience.
Flannery, B. P., Kheshgi, H., Marland, G., & MacCracken, M. C. (1997). Geoengineering climate. In R. G. Watts (Ed.), *Engineering response to global climate change* (pp. 379–427). Boca Raton, FL: Lewis.
Fleming, J. R. (2004). Fixing the climate: Military and civilian schemes for cloud seeding and climate engineering. In L. Rosner (Ed.), *The technological fix: How people use technology to create and solve problems* (pp. x–xx). London: Taylor & Francis.
Fleming, J. R. (2006). The pathological history of weather and climate modification: Three cycles of promise and hype. *Historical Studies in the Physical and Biological Sciences, 37*(1), 3–25.
Fleming, J. R. (2010a). *Fixing the sky: The checkered history of weather and climate control*. New York: Columbia University Press.
Fleming, J. R. (2010b). Planetary-scale fieldwork: Harry Wexler on the possibilities of ozone depletion and climate control. In J. Vetter (Ed.), *Knowing global environments: New historical perspectives on the field sciences* (pp. 190–211). New Brunswick, NJ: Rutgers University Press.
Fletcher, J. O. (1968). *Changing climate*. Santa Monica, CA: RAND Corporation.
Forman, P. (1987). Behind quantum electronics: National security as basis for physical research in the United States, 1940–60. *Historical Studies in the Physical and Biological Sciences, 18*(1), 149–229.
Fukuyama, F. (1989). The end of history? *The National Interest, 16*(Summer), 3–18.
Gilman, N. (2003). *Mandarins of the future: Modernization theory in Cold War America*. Baltimore, MD: Johns Hopkins University Press.

Goodchild, P. (2004). *Edward Teller: The real Dr Strangelove*. London: Weidenfeld & Nicholson.
Goodell, J. (2006, November 3). Can Dr. Evil save the world? *Rollingstone.com*.
Haas, P. M. (1992). Introduction: Epistemic communities and international policy coordination. *International Organization, 46*, 1–35.
Harper, K. (2008). Climate control: United States weather modification in the cold war and beyond. *Endeavour, 32*(1), 20–26.
Harper, K. C., & Doel, R. E. (2010). Environmental diplomacy in the Cold War: Weather control, the United States, and India, 1966–1967. In J. R. McNeill & C. R. Unger (Eds.), *Environmental histories of the Cold War* (pp. 115–138). Cambridge: Cambridge University Press.
Havel, V. (1992). Speech to World Economic Forum. Retrieved January 28, 2019, from https://web.archive.org/web/20131203150151/http://vaclavhavel.cz/showtrans.php?cat=projevy&val=265_aj_projevy.html&typ=HTML
Hess, W. N. (Ed.). (1974). *Weather and climate modification*. New York: Wiley-Interscience.
Hünemörder, K. (2010). Environmental crisis and soft politics: Détente and the global environment, 1968–75. In J. R. McNeill & C. R. Unger (Eds.), *Environmental histories of the Cold War* (pp. 257–276). Cambridge: Cambridge University Press.
IPCC (Intergovernmental Panel on Climate Change). (1995a). *Second Assessment Climate Change 1995, synthesis report*. Geneva: IPCC.
IPCC (Intergovernmental Panel on Climate Change). (1995b). *Impacts, adaptations and mitigation of climate change: Scientific-technical analyses*. Working Group 2 report. Geneva: IPCC.
IPCC (Intergovernmental Panel on Climate Change). (2000). *Third Assessment Report*. Working Group 3, Mitigation. Geneva: IPCC.
IPCC (Intergovernmental Panel on Climate Change). (2012). *Meeting report of the Intergovernmental Panel on Climate Change expert meeting on geoengineering* (O. Edenhofer, R. Pichs-Madruga, Y. Sokona, C. Field, V. Barros, T. F. Stocker, Q. Dahe, J. Minx, K. Mach, G.-K. Plattner, S. Schlömer, G. Hansen, & M. Mastrandrea, Eds.). IPCC Working Group III Technical Support Unit, Potsdam Institute for Climate Impact Research. Geneva: IPCC.
Josephson, P. (2011). Technology and the conquest of the Arctic. *The Russian Review, 70*(3), 419–439.
Keith, D. W. (2000). Geoengineering the climate: History and prospect. *Annual Review of Energy & Environment, 25*, 245–284.

Keith, D. W., & Dowlatabadi, H. (1992). Taking geoengineering seriously. *Eos, Transactions, American Geophysical Union, 73*(27), 289–296.

Kellogg, W. W., & Schneider, S. H. (1974). Climate stabilization: For better or for worse? *Science, 186*(4170), 1163–1172.

Kennedy, J. F. (1961, September 25). Address at U.N. General Assembly. Retrieved January 7, 2019, from http://www.jfklibrary.org/Asset-Viewer/DOPIN64xJUGRKgdHJ9NfgQ.aspx

Kirsch, S., & Mitchell, D. (1998). Earth-moving as the 'Measure of Man': Edward Teller, geographical engineering, and the matter of progress. *Social Text, 54*(16), 101–134.

Litfin, K. T. (1995). Framing science: Precautionary discourse and the ozone treaties. *Millennium: Journal of International Studies, 24*(2), 251–277.

MacCracken, M. C. (1991). *Geoengineering the climate*. Paper prepared for 'Workshop on the Engineering Response to Global Climate Change', Palm Coast, FL, 1–6 June. Lawrence Livermore National Laboratory.

MacCracken, M. C. (2006). Geoengineering: Worthy of cautious evaluation? *Climatic Change, 77*(3–4), 235–243.

Macrae, N. (1992). *John von Neumann: The scientific genius who pioneered the modern computer, game theory, nuclear deterrence, and much more*. New York: Pantheon Books.

Marchetti, C. (1977). On geoengineering and the CO_2 problem. *Climatic Change, 1*(1), 59–68.

Marland, G. (1996). Could we/should we engineer the Earth's climate? *Climatic Change, 33*(3), 275–278.

Masco, J. (2010). Bad weather: On planetary crisis. *Social Studies of Science, 40*(1), 7–40.

McNeill, J. R., & Unger, C. R. (Eds.). (2010). *Environmental histories of the Cold War*. Cambridge: Cambridge University Press.

Meadows, D. H., Meadows, D. L., Randers, J., & Behrens III, W. W. (1972). *The limits to growth: A report for the club of Rome's project on the predicament of mankind*. New York: Universe Books.

Miller, C. (2004). Climate science and the making of a global political order. In S. Jasanoff (Ed.), *States of knowledge: The co-production of science and social order* (pp. 46–66). London and New York: Routledge.

Milman, O. (2016, April 14). Oil industry knew of 'serious' climate concerns more than 45 years ago. *The Guardian*. Retrieved January 7, 2019, from https://www.theguardian.com/business/2016/apr/13/climate-change-oil-industry-environment-warning-1968

Milonakis, D., & Fine, B. (2009). *From political economy to economics: Method, the social and the historical in the evolution of economic theory.* Abingdon and Oxford: Routledge.

NAS (National Academy of Sciences). (1992). Geoengineering. In *Policy implications of greenhouse warming: Mitigation, adaptation and the science base.* Washington, DC: National Academies Press.

NAS-NRC (National Academy of Sciences/National Research Council Committee on Atmospheric Sciences). (1971). *The atmospheric sciences and man's needs: Priorities for the future.* Washington, DC: National Academy of Sciences.

New York Times. (1945, December 9). Julian Huxley pictures the more spectacular possibilities that lie in atomic power, p. 77.

Oreskes, N., & Conway, E. M. (2010). *Merchants of doubt: How a handful of scientists obscured the truth on issues from tobacco smoke to global warming.* New York: Bloomsbury Press.

Oreskes, N., & Krige, J. (Eds.). (2014). *Science and technology in the global Cold War.* Cambridge, MA: MIT Press.

Philander, S. G. (Ed.). (2008). *Encyclopedia of global warming and climate change.* Thousand Oaks, CA: SAGE.

Pielke, R., Prins, G., Rayner, S., & Sarewitz, D. (2007). Climate Change 2007: Lifting the taboo on adaptation. *Nature, 445*(7128), 597.

Poole, R. (2008). *Earthrise: How man first saw the Earth.* New Haven, CT: Yale University Press.

Robinson, E., & Robbins, R. C. (1968). *Sources, abundance, and fate of gaseous atmospheric pollutants.* Final report and supplement. Menlo Park, CA: Stanford Research Institute.

Robinson, J. (2004). Squaring the circle? Some thoughts on the idea of sustainable development. *Ecological Economics, 48*(4), 369–384.

Rusin, N. P., & Flit, L.A. (1962). *Methods of climate control.* Translated from Russian 1964. TT 64–21333. Washington, DC: Department of Commerce, Office of Technical Services, Joint Publications Research Services.

Sachs, W. (Ed.). (2010). *The development dictionary: A guide to knowledge as power* (2nd ed.). London and New York: Zed Books.

Schelling, T. C. (1996). The economic diplomacy of geoengineering. *Climatic Change, 33*(3), 303–307.

Schneider, S. (1996). Geoengineering, could—Or should—We do it? *Climatic Change, 33*(3), 291–302.

Schneider, S. (2001). Earth systems engineering and management. *Nature, 409*(6818), 417–421.

Shaw, D. J. B. (2015). Mastering nature through science: Soviet geographers and the Great Stalin Plan for the transformation of nature, 1948–53. *The Slavonic and East European Review (Special Issue: Conceptualizing and Utilizing the Natural Environment: Critical Reflections from Imperial and Soviet Russia), 93*(1), 120–146.
SMIC (Study of Man's Impact on Climate). (1971). *Inadvertent climate modification: Report of the Study of Man's Impact on Climate (SMIC)*. Cambridge, MA: MIT Press.
Steger, M. B. (2009). *Globalisms: The great ideological struggle of the twenty-first century* (3rd ed.). Lanham, MD: Rowman & Littlefield.
Teller, E. (1997, October 17). Sunscreen for planet earth. *Wall Street Journal*. Retrieved January 7, 2019, from http://www.hoover.org/research/sunscreen-planet-earth-0
Teller, E. with Brown, A. (1962). *The legacy of Hiroshima*. London: Macmillan.
Teller E., Wood, L., & Hyde, R. (1997). *Global warming and ice ages: I. Prospects for physics based modulation of global change*. UCRL-JC-128157. Livermore, CA: Lawrence Livermore National Laboratory.
Uekötter, F. (2010). The end of the Cold War: A turning point in environmental history? In J. R. McNeill & C. R. Unger (Eds.), *Environmental histories of the Cold War* (pp. 343–352). Cambridge: Cambridge University Press.
Wallerstein, I. (2002, July). The Eagle has crash landed. *Foreign Policy*. Retrieved from www.foreignpolicy.com/issue_julyaug_2002/wallerstein.html
Weart, S. R. (2008). *The discovery of global warming*. Cambridge, MA: Harvard University Press.
Westad, O. A. (2007). *The global Cold War*. Cambridge: Cambridge University Press.
White House. (1965). *Restoring the quality of our environment: Report of The Environmental Pollution Panel, President's Science Advisory Committee*, November.
Zikeev, N. T., & Doumani, G. A. (1967). *Weather modification in the Soviet Union, 1945–1966: A selected annotated bibliography*. Washington, DC: Library of Congress Science and Technology Division [Supt. of Docs., U.S. Govt. Print. Off.].

3

The Re-emergence of Solar Geoengineering

As the decade of the 1990s progressed the shine was coming off earlier suggestions of the 'end of history'. The promotion of democracy, often by selectively imposing governance-related conditionalities, bred resentment in the global South and highlighted their dependent status. Moves towards privatisation and fiscal de-regulation sometimes led to economic turbulence: such as the collapse of the Eastern bloc economies and the looting of public assets by a new elite; or the economic meltdowns and debt crises across South-East Asia in 1997; or the devastation and conflict generated in Africa in the wake of 'structural adjustment programmes'. Prioritising the pursuit of GDP growth became an article of faith in most countries and in all multilateral economic institutions. Large swathes of the emerging Southern elite aspired to imitate or exceed US-style consumption patterns, and visibly so. From an environmental perspective it is hard to overstate the magnitude of these developments and the extent of ecological degradation which has followed. As Christoff and Eckersley argue: "… the contemporary dominant form of *economic* globalization (capitalism) serves as the main intensifier of environmental degradation" (2013: 13, emphasis in original).

There were many multilateral environmental initiatives and agreements during this period, often innovative ones, but little decline in the pace and intensity of ecological destruction. The central concept of 'sustainable development' became increasingly emptied of meaning. As Bernstein and Ivanova put it, by the mid-2000s:

> ... the promise of sustainable development remain[ed] largely unfulfilled and environmental governance ha[d] evolved largely in conformity with the changing demands of a hyper-liberal global political economy, rather than vice versa. (2008: 162)

Further, the securitisation of the globalisation project, especially after the attacks of September 11th 2001, meant that tackling environmental issues dropped further down the policy agenda, behind economic growth and waging war on Jihadist globalism. Importantly for our analysis, when environmental issues were considered post-2001, this was often in a securitised mode, such as the framing of climate change as a security concern (Dalby 2009).

The initial optimism in the early 1990s, that history had ended and that new forms of global regulation might help address ecological damage, was soon supplanted by a more pessimistic mood. In short, we find from around the turn of the millennium, a pessimistic, increasingly securitised, market globalism under US hegemony, one manifestly incapable of addressing environmental degradation and climate change. This was the broad social, economic, ideational and geo-political context within which geoengineering was to re-emerge.

Climate Urgency and Emergency

The ways in which climate change was imagined and understood in the 'developed world' in the first years of the 2000s was also changing.[1] Here I simply sketch the contours of the time relevant to geoengineering's re-emergence.

---

[1] Here I limit myself to the 'developed world' since it is there, mainly in the United States and to a lesser extent the UK, that the idea of geoengineering re-emerges.

3 The Re-emergence of Solar Geoengineering 77

From the start of the millennium, the signs were clear that the United States, the world's largest greenhouse gas emitter, had neither the appetite nor the desire to cut emissions. This was not encouraging for anyone wanting to tackle climate change through mitigation. After all, the Kyoto Protocol had set the modest goal of reducing greenhouse gas emissions, by the year 2012, to 5% below 1990 levels.

In June 2001 President George W. Bush announced that the United States would not sign up to the Kyoto Protocol to which it had previously agreed. Reasons given for rejecting this 'flawed treaty' included that it imposed no obligations on developing countries like China and India, that there was insufficient knowledge about global warming, that the Kyoto targets were "arbitrary and not based upon science", and that "complying with those mandates would have a negative economic impact" (White House 2001). Instead, a target for improving emission intensity was announced, one that would allow emissions to increase, but at a slower pace. Pulling out of Kyoto did not come as a surprise. In his second presidential debate with Al Gore in 2000, Bush had made his approach clear:

> I'll tell you one thing I'm not going to do is I'm not going to let the United States carry the burden for cleaning up the world's air like the Kyoto treaty would have done. (cited in Singer 2002: 27)[2]

The 2002 Conference of the Parties (CoP) meeting in New Delhi, in a sop to the US, saw references to the Kyoto Protocol dropped from the draft statement. Russia, whose ratification of the treaty was essential in order to bring it into effect, was hesitating. The 2003 CoP in Milan saw the ongoing instability in post-communist Russia on display. The deputy Economy Minister said Russia was moving towards ratification, and in practice Russian emissions had dropped sharply as its economy went into a tailspin. But he was flatly contradicted by President Putin's economic advisor who said "[i]t is impossible to undertake responsibilities that place serious limits on the country's growth" (Kirby 2003).

---

[2] His father President George Bush Snr. had spoken along similar lines at the Rio Earth Summit in 1992 when he said "the American lifestyle is not up for negotiation" (cited in Singer 2002: 2).

A review of the year 2003, published in *Nature*, noted that "[e]nvironmentalists may remember 2003 as the year in which the Kyoto Protocol died" and that "the international community's first attempt at tackling climate change is in terminal decline" (Schiermeier 2003: 756). Some signs of the times: an unusually intense European heatwave in the summer of 2003 resulted in an estimated 15,000 deaths in August in France alone (Robine et al. 2008). Shortly thereafter major wildfires swept across Southern California. In August 2005 Hurricane Katrina struck New Orleans to devastating effect. Examples of extreme weather such as these suggested the severity of global warming already being felt, notwithstanding the standard scientific response at the time that no particular weather event could be attributed to climate change. Events such as these helped shape an emerging narrative of climate change as the biggest environmental problem of all, one which posed an existential threat to humanity. Talk of needing to go onto a 'war' footing, and for Churchillian-style leadership, starts to emerge in some accounts of this time (see Dibley and Neilson 2010 for an account of this).[3] In general this framing stressed crisis, urgency and emergency, and called for action in the face of impending catastrophe. An example can be found in a talk at Yale University in April 2004 by former US Vice-President Al Gore, entitled 'The Climate Emergency':

> The title I chose for this speech is not a misprint... this is a crisis with an unusual sense of urgency attached to it, and we should see it as an emergency (p. 154). ... [T]his is happening right now. And it carries with it, unless we do something, catastrophic consequences for all civilization. (Gore 2004: 157)

Gore's influential 2006 movie, *An Inconvenient Truth*, was still to follow and would continue this theme. But it faced a competing narrative which rejected any suggestion of a crisis, labelling such talk alarmist and accusing its proponents of exaggeration and, occasionally, conspiracy.

---

[3] It is symptomatic of this period of market globalism that a leading billionaire airline entrepreneur and celebrity, Richard Branson, is selected to give a keynote address on climate change to the UN General Assembly. In it he calls for an "Environmental War Room" to be established and for mankind to "regulate the Earth's temperature" (Branson 2008: n.p.).

The competing narrative, dominant within the Bush presidency, stressed the naturalness of climate variations, the uncertainty of the science, and that climate impacts were not imminent but in the future. Apart from those who denied that climate change was an issue at all, they either accepted modest climate measures or actively called for inaction.

We now know, as Brysse et al. have shown, that the "official" findings of climate science during this period probably erred on the side of caution rather than alarmism when it came to making predictions. "Scientists, particularly acting in the context of large assessments, may have underestimated the magnitude and rate of expected impacts of anthropogenic climate change" (Brysse et al. 2013: 335): an example being estimates of sea-level rise. If anything, the scientists were understating the severity of the data… and they often knew it.

At the turn of the millennium a cautious policy 'consensus' about climate change was becoming apparent (Oreskes 2004) and, most significantly, was being communicated to policymakers, leading politicians and even the general public. The IPCC's Third Assessment Report published in 2001 included, for the first time, a separate synthesis and summary for policymakers and paid greater attention than previously to presentation. The emerging 'consensus' crystallised around a number of propositions. First, that global warming was occurring: as the IPCC's Third Assessment Report put it, the 1990s was the warmest decade, and 1998 the warmest year, since instrumental records began in 1861 (2001: 4). Second, that the cause was primarily anthropogenic, with "new and stronger evidence that most of the warming observed over the last 50 years is attributable to human activities" (2001: 5–7). Contrast this with the 1995 Second Assessment Report with its more modest claim that "the balance of evidence suggests that there is a discernible human influence on global climate" (IPCC 1995a: 22). Third, that the effects of increasing climate change would be mainly adverse, that the impacts would "fall disproportionately upon developing countries and the poor persons within all countries", including that it would "exacerbate water shortages in many water-scarce areas of the world" (IPCC 2001: 9–12). Contrast this with the Second Assessment Report which, although it contained some similar conclusions in the small print, headlined the relevant section "Impacts are difficult to quantify and existing studies are limited" (1995a: 29). A fourth conclusion was that significant emissions cuts were needed, although the modelling of various scenarios produced very different estimations of what was required, and with large error ranges. The take-home message being conveyed in the Third Assessment Report was that 'it's happening, it's us, it's possibly serious, and emission cuts are needed'.

In short, the climate context within which geoengineering was to re-emerge included the 'consensus' science expressing growing concern, mitigation responses to climate change which were underwhelming in scale and scope, and the 'climate problem' increasingly being articulated as both a global crisis and a civilizational emergency.[4]

Feeding into the emergency/crisis discourse, during this period a number of terms from the climate sciences were becoming more commonplace in the adjacent climate policy world. Two bundles of concepts are especially important for consideration of solar geoengineering (SGE). First, ideas of 'non-linearity', 'hysteresis' and the possibility of abrupt and hard-to-predict changes in the climate system. Most dramatically presented as 'tipping points', these were understood as critical thresholds in non-linear systems beyond which changes might be unleashed which could not be reined back in. Elements of the Earth system, such as the West Antarctic Ice Sheet or the release of methane from melting permafrost, were understood to be at particular risk of 'tipping'. Second, and related, 'inertia' and the concern that, once set into motion, many climate changes (such as sea level rise or ocean acidification) would lumber along, unfolding over centuries, even millennia, and be effectively 'irreversible' even with emission cuts. In the field of public policy, growing awareness of the complex and 'coupled' system dynamics associated with climate change, led to the revival and reworking of concepts such as "wicked problems" (Rittel and Webber 1973).

James Hansen, amongst the most highly-regarded climate scientists of his generation, summarised the situation as he saw it in a speech to the peak event of his disciplinary community, the conference of the American Geophysical Community (AGU) in December 2005:

---

[4] This, in effect, converted a range of existing environmental concerns—such as extinction, biodiversity, toxicity and waste—into second-order environmental issues. See for example Crist (2007), who examines the implications of this.

The Earth's climate is nearing, but has not passed, a tipping point beyond which it will be impossible to avoid climate change with far-ranging undesirable consequences. These include not only the loss of the Arctic as we know it, with all that implies for wildlife and indigenous peoples, but losses on a much vaster scale due to rising seas… This grim scenario can be halted if the growth of greenhouse gas emissions is slowed in the first quarter of this century. (cited in Bowen 2008: 4)

In short, the climate context, scientifically and discursively, was one which stressed urgency, seriousness, the need for determined leadership, and a sense that the time to act was running out. Whilst there were exceptions, as we shall see shortly, in general SGE re-emerges closely attached to climate emergency and crisis narratives.

Geoengineering's Re-emergence

In late 2001 the US National Research Council convened a committee to examine and report on weather modification. There had been a gap of almost three decades since the previous reports in 1964, 1966 and 1973. Its report, published two years later, carries the marks of the 'taboo' period when it notes that weather modification generates "serious feelings of ambivalence" in the scientific community and the support agencies (NRC 2003: iv); and when it partly disregards its brief which "explicitly excluded consideration of the complex social and legal issues" (p. viii); or when it stresses that the results of improved scientific knowledge of atmospheric processes are "… as likely to lead to viable weather modification methodologies as they are to indicate that intentional modification of a weather system is neither currently possible nor desirable" (p. vii). The committee's brief included examining research directions relevant to "reduction in severe weather hazards" and specifically covered both "localized weather modification" and "large-scale weather and climate patterns". The significance of the 2003 report lies not in what it says but in its silences. It contained *no* discussion, *no* mention even, of geoengineering. The topic was still effectively taboo.

Around this time, geoengineering is considered in a scenario prepared for the Pentagon in October 2003 by leading scenario specialists Peter Schwartz and Doug Randall. Whilst the document is not an official

Pentagon one, it is of interest as suggestive of the types of planning scenarios under consideration at that time. Schwartz and Randall's (2003) stated aim is to "imagine the unthinkable", and present a "plausible scenario" of abrupt climate change following a sharp slowdown in the North Atlantic thermohaline conveyor. This would be associated with changes to weather patterns across much of the world (both warming and cooling), as well as food shortages, water crises and warfare. They conclude with seven recommendations, one of which is to "explore geo-engineering options that control the climate", although they caution that such an action "would have the potential to exacerbate conflict among nations" (2003: 22). Geoengineering is mentioned, but in an explicitly 'outside-the-box' and non-official document.

The 2004 Tyndall Centre Gathering

A few months later, in January 2004, the Tyndall Centre for Climate Change Research and the [University of] Cambridge-MIT Institute convened a special joint Symposium on 'Macro-Engineering Options for Climate Change Management and Mitigation'. The symposium report starts by outlining the background: "many people" recognise it will be "extremely difficult" to reduce emissions, especially if combined with per capita convergence requirements requiring greater cuts by developed countries. "The urgency of implementing climate change management [means] more innovative approaches to … mitigation" will likely be needed (Tyndall Centre 2004: 1). Nuclear fission and fusion will not make up the shortfall in the medium-term and so "macro-engineering options for climate change management" (i.e. geoengineering) need to be discussed and evaluated before they can be "seriously considered for implementation" (p. 1).

The symposium covered a range of proposed technologies including carbon capture and storage (CCS), ocean fertilisation, cloud stimulation and, most relevant for our purposes, stratospheric aerosols. It is worth quoting at some length the report section entitled 'Philosophy' as this helps in understanding the imaginary and assumptions then prevalent. It also makes clear that what was being discussed was still 'taboo', and that there was a need to rationalise breaking that taboo.

Although most of the macro-engineering approaches identified so far are *not currently in the mainstream* thinking in relation to climate policy, the mere fact that they have been conceived and proposed places an obligation on engineers, economists, and environmental and social scientists, working together, to explore their feasibility and evaluate their consequences and their wider implications. At the very least, such options may need to be considered as *emergency policy options* in the event of greater adverse climate change impacts than expected, or less effective carbon reduction measures than anticipated. The process of exploration, evaluation, development and (eventually) pre-operational implementation of such approaches should be regarded as at least an *insurance* against these eventualities.

Many of these possible options are *highly speculative* at present, and some may even appear to be quite *crazy*. However, that is precisely why they should be evaluated (and if necessary dismissed) as soon as possible. Otherwise, politicians may seek to use them as "Magic Bullets" either to postpone action, or as prospective solutions for actual implementation, once it becomes clear that the *mitigation of climate change is going to be a major and very difficult task indeed…*

The symposium therefore aimed to

- Consider all approaches identified, objectively, and without preconceptions
- Engage in an open, unbiased, and visionary but still concrete discussion
- Disregard potential pressures in relation to *political correctness.*
- Employ a very wide range of criteria for a preliminary evaluation
(Tyndall Centre 2004: 2, all emphases my own)

One can see that geoengineering here is tied to 'emergency', 'urgency', and seen as an 'insurance' policy in a context where climate mitigation seems impossible. There is awareness that these ideas may be seen as 'crazy', and a self-perception by the attendees that they are fearless, even 'visionary', straight-talking and unbowed by 'political correctness' (presumably referring to the *de-facto* 'taboo' on geoengineering then operative).

Only 34 invited attendees from North America, the UK and elsewhere in Europe participated. The Stratospheric Aerosol discussion was led by Lowell Wood, a long-standing proponent for 'albedo modification' and explicit political conservative.[5] Indeed, the brief symposium report, when discussing 'Next Steps' makes the point that "the debate could now usefully be re-envigorated [sic] in North America, where a way is needed to build further on the 1992 NAS report" (chapter 28), of which Wood had been a co-author. The report envisages the need for more basic research, large-scale pilot experiments, and large-scale tests (Tyndall Centre 2004: 5). It cautions against using research as a reason to delay implementation of known existing solutions. The symposium report notes that all participants agreed to promote the idea that 'macroengineering options' "… needed to be brought into the main-stream of the debate on possible responses to climate change" (p. 6).

Wood's paper to the Cambridge-MIT Institute symposium is not available. But his thinking at the time can be found in a 2002 paper for the National Academy of Engineering that he co-authored with Teller and Hyde, his colleagues at Lawrence Livermore. This paper largely reprises the calculations done for the 1992 NAS study and argues that stratospheric aerosols are a cheap and effective solution: entailing "… expenditures of no more than $1 B[n]/year, commencing not much sooner than a half-century hence, even in worst-case scenarios" (Teller et al. 2002: 8). It makes its normative case more transparently than in the 1992 NAS study, and one can hear his voice as advisor to George W. Bush coming through. The authors make the unusual argument that solar geoengineering's cheapness means that "technical management of radiative forcing of the Earth's fluid envelopes, not administrative management of gaseous inputs to the atmosphere, is the path mandated by the pertinent provisions of the UN Framework Convention on Climate Change" (p. 8).[6] The technique is seen as doubly cheap as not

---

[5] We have already encountered Lowell Wood in the previous Chapter as a close colleague of Edward Teller, as a co-author of the largely ignored geoengineering chapter in the 1992 NAS study on climate change, and as one of the advisors in the unsuccessful attempt by George W. Bush's climate advisors to put geoengineering on the policy agenda.

[6] Wood's argument, in his and his co-author's words, is: "… *if* you're inclined to subscribe to the Rio Framework Convention's directive that mitigation of global warming should be effected in the 'lowest possible cost' manner—*whether or not* you believe that the Earth is indeed warming significantly above-and beyond natural rates, and *whether or not* you believe that human activities are largely responsible for such warming, and *whether or not* you believe that problems likely to have significant impacts only a century hence should be addressed with current technological ways-&-means rather than be deferred for obviating with more advanced means—then you will

only is its cost low, but by allowing CO_2 concentrations to rise, plant fertilisation is enhanced, thereby enabling humankind to get around "the basic food production challenge of the twenty first century" (p. 8). In essence, solar geoengineering is presented as a better alternative, *if* climate change is a problem, to mitigation via emissions reductions, and with added 'development' benefits. This is not about 'urgency' or 'emergency'.

This recapitulates the thinking of Wood's mentor Edward Teller. In a 1997 article in the *Wall Street Journal* Teller had called for "a sunscreen for planet Earth", for a problem he clearly thought had been hyped:

> [F]or some reason, [the geoengineering] option isn't as fashionable as all-out war on fossil fuels and the people who use them. Yet if the politics of global warming require that 'something must be done' while we still don't know whether anything really needs to be done—let alone what exactly—let us play to our uniquely American strengths in innovation and technology to offset any global warming by the least costly means possible. … Injecting sunlight-scattering particles into the stratosphere appears to be a promising approach. Why not do that? (Teller 1997)

Echoes of the 'Mastery' imaginary, discussed in the previous Chapter, are clearly evident. And the trope—geoengineering as an alternative plan for tackling climate change—can still be found today. It remains attractive to both the politically conservative, since it explicitly rules out tackling the dominant energy and economic order, and also to the climate 'sceptical', since it provides a backstop "in case the science turns out to have been right" (for examples see Lomborg interviewed by Dickinson 2010; Gingrich 2008; Levitt and Dubner 2009). But it was a perspective which had failed to gain traction in the 1990s and was unlikely to do so in the 2000s if argued in this way, even as the seriousness of the climate condition became more, rather than less, concerning.

If geoengineering was to become a part of mainstream climate policy consideration it would need to be detached from climate inactivist thinking and attached to the mainstream scientific narrative that stressed the magnitude of the anthropogenic climate issue, not its uncertainty. And it

---

necessarily prefer active technical management of radiation forcing of the Earth to administrative management of greenhouse gas inputs to the Earth's atmosphere, for the practical reasons sketched in the foregoing" (Teller et al. 2002: 6, emphases in original).

needed a proponent with greater credibility and less ideological baggage than Teller or the Lawrence Livermore circle. This happened in 2006.

Crutzen and the Lifting of the Taboo

The entry of geoengineering into mainstream *policy* consideration can probably be dated to a 2006 editorial by Paul Crutzen in the journal *Climatic Change* entitled 'Albedo Enhancement by Stratospheric Sulfur Injections: a contribution to resolve a policy dilemma?' (Crutzen 2006). Crutzen was a prominent Dutch atmospheric chemist and climate scientist, Nobel prize winner for chemistry for his work linked to the ozone hole, and influential thereby in shaping one of the few international environmental treaties (the Montreal protocol) often regarded as a model for climate change regulation to follow, as well as populariser of the idea of 'nuclear winter' and the idea of the 'Anthropocene'.

In the article Crutzen outlined the science indicating the seriousness of the warming projections contained in the models. Relying on the Pinatubo eruption as proof of concept, he drew on two key arguments. Firstly, concern that initiatives to reduce smog and pollution, for health and ecological reasons, may paradoxically result in a potentially catastrophic spike in temperatures and the need "to combat potentially drastic climate heating" (2006: 216). Secondly, that CO_2 emissions have continued to rise when "by far the preferred way" was that they fall by 60–80%, and that the "failure of mitigation" showed that the political response to the climate problem had been inadequate (pp. 211–12).

"[A]lthough by far not the best solution, the usefulness of artificially enhancing earth's albedo and thereby cooling climate by adding sunlight reflecting aerosol in the stratosphere … might again be explored and debated" (p. 212). Crutzen stressed that: "Importantly, its possibility should not be used to justify inadequate climate policies, but merely to create a possibility to combat potentially drastic climate heating" (p. 216). The crux of his argument was:

> Given the grossly disappointing international political response to the required greenhouse gas emissions, and further considering some drastic results of recent studies [that the prognosis is worse than previously

believed], ... research on the feasibility and environmental consequences of climate engineering ... which might need to be deployed in future, should not be tabooed. (p. 214)

Van Hemert has shown how the scientific basis of Crutzen's argument was a highly-contestable, back-of-the-envelope calculation, which included arguable claims about this speculative technology's cooling effect and cheapness (2017).

Prior to publication, according to Ralph Cicerone, a climate scientist and the journal editor, a number of his scientific colleagues appealed to him not to publish (Cicerone 2006: 221). Unusually for an academic journal, Crutzen's contribution ended up being published alongside a number of responses, some critical of his position and stressing the many flaws in his argument, including the regionally uneven mixture of warming, cooling and drying that followed Pinatubo (for example Kiehl 2006). Once pushed out of the closet and into the public domain, geoengineering could not be re-closeted. Not that the idea had been secret. As we have seen, there was an existing, albeit limited, scientific literature on, or relevant to, geoengineering. But this mainly emanated from scientists associated with earlier and discredited visions of mastery. And there were pockets of support for the idea elsewhere, such as from economists Nordhaus (1994) and Schelling (1996) in the wake of the 1992 NAS report. But Crutzen's stature, the deliberateness of the gesture and the forum chosen, ensured the idea of geoengineering entered into the domain of respectable climate policy discussion. The taboo on discussing geoengineering was lifted, at least in the minds of a number of influential climate scientists.

The NASA Workshop

Shortly thereafter, in November 2006, the NASA Ames Research Center, together with the Carnegie Institution sponsored an expert workshop on the use of what they called 'solar radiation management'. The focus was on what technologies might be used, their effectiveness and unintended consequences. It had to acknowledge the contentious nature of the idea. The workshop report, 'Managing Solar Radiation' stated that participant views on "... the circumstances under which solar radiation management should be deployed ...", ranged from "(i) never, (ii) only in the event of an imminent climate catastrophe, (iii) as part of a transition to a low-carbon-emission economy, and (iv) in lieu of strong reductions in greenhouse gas emissions" (Lane et al. 2007: vi).[7]

In the report's climate policy section, option (iv) is largely rolled into option (iii). Revealingly, option (i) makes no further appearance in the report. The positions from only a year earlier have been reversed—thoughts of 'taboo' have been assigned to the margins by the authors. Accordingly, the report's climate policy section sets out two "rival strategic visions":

> One of these, which might be called the parachute strategy, would foresee deployment only in the event of a climate change emergency. The second, preemptive deployment strategy, would implement solar radiation management technologies as soon as research firmly established their safety and efficacy. (Lane et al. 2007: 11)

The emergency vision was understood to be "politically straightforward"—'this would be an emergency', after all—seems to have been the assumption! The technology would be developed and then put on the shelf. The emergency view, as we shall see in Chapter 5, is far from being "straightforward", either politically or scientifically.

The pre-emptive vision was understood as "a temporary measure to buy time for emission reductions" and develop new technologies.[8] It

---

[7] Prior to lifting the taboo the last three views would barely have received a hearing! The third view was also unusual and suggested the pre-emptive use of solar geoengineering as part of a suite of policy interventions into climate. It is associated with Tom Wigley (2006), who also attended this workshop. The related proposal of experimenting in the atmosphere, starting small and then "cautiously" scaling up any intervention, was also voiced (Lane et al. 2007: 6). This is the argument most popular amongst SGE's proponents today.

[8] Not discussed is how the conception of the 'temporary' nature of any intervention might be reconciled with SGE being a long-term commitment. The 'termination effect' as it is now termed, is clearly outlined in the later 'official' literature (for example NRC 2015: 36) but was perhaps not fully acknowledged at the time.

could "… be consistent with an *economically efficient* climate policy… postponing the deepest emission cuts until cheaper abatement technology is available" (p. 11, my emphasis).

2006 can be seen as a turning point. It was the moment when geoengineering became a respectable idea at least for inclusion in expert discussions on climate policy. Whilst the Tyndall symposium of 2004 imagined its proceedings as transgressive, politically incorrect and 'crazy', the NASA Ames symposium of 2006 thought of itself as mainstream and responsible and focussed not on whether to geoengineer but on the circumstances under which to do so, and the knowledge needed.

Geoengineering's emerging respectability was a turn away from Taboo. But it was not, generally speaking, a move towards its unconditional embrace. Even amongst its proponents SGE was widely seen as, to use Kintisch's phrase, "a bad idea whose time has come" (2010: 13). And its move into the mainstream as a policy option provoked, as we shall soon see, extensive pushback against the idea itself and also prompted calls for the regulation of research in the area and for restrictions on any suggestion of experimentation with the atmosphere (or the oceans).

Geoengineering's becoming mainstream, at least in climate policy discussions, is further evidenced by the veritable flood of academic research, commentary, and popular accounts, across a range of disciplines, about geoengineering from 2006 (Belter and Seidel 2013; Oldham et al. 2014).

Respectable but Not Embraced

As I explore elsewhere, SGE generates a wide range of responses ranging from enthusiasm to revulsion and much in between. Here I focus on geoengineering's 'official' status and how it has been received in the institutions and policy circles which have the power to shape climate policy. This is the realm of policy speaking with power. Two terrains are key: the United States, since it is hard to imagine geoengineering materialising as a practice without the explicit or implicit approval of the United States; and the UNFCCC/IPCC as the dominant global space where climate expertise and climate policymaking meet. Since the lifting of the taboo, geoengineering has become increasingly visible in official and authoritative climate policy documents and in a range of publicly-funded initiatives.

In the United States there are reports associated with committees of Congress (for example USHCST 2009, 2010) and Senate (Schnare 2007), and detailed reports produced by the Government Accountability Office (GAO 2010, 2011). Other relevant official and 'close to power' studies include the NASA-Ames study already mentioned (Lane et al. 2007), a report commissioned by the RAND Corporation on governing geoengineering research (Lempert and Prosnitz 2011), and reports by or for the Council on Foreign Relations (Ricke et al. 2008), and the Bipartisan Policy Center (2011). These are close to the military, the foreign policy and the political establishment, respectively. But the most comprehensive work has been done by the National Research Council, in a project largely sponsored by the US intelligence establishment. This resulted in a major study, the most comprehensive and thorough to date, entitled *Climate Intervention: Reflecting Sunlight to Cool the Earth* (NRC 2015).[9] This report could not be said to endorse solar geoengineering, indeed it could be read as implicitly critical of the idea… at least for now. But it did not take it off the table as an option. Beyond these reports, the US state of Arizona is the specified location for early experimentation: the Stratospheric Controlled Perturbation Experiment (SCoPEx) run out of Harvard University and planned for 2018 but not yet embarked upon at the time of writing.

The IPCC is also paying increasing attention to solar geoengineering. Compared to the first four Assessment Reports—in 1990, 1995, 2001 and 2007—the Fifth Assessment Report of 2013 significantly increased the amount of coverage and comment devoted to geoengineering: the topic is covered in all three working groups and in the synthesis report and the summary for policymakers. It was preceded by a special global expert meeting devoted to the subject (IPCC 2012). The shift in tone in the Fifth Assessment Report is subtle but noticeable. The synthesis report remains cautious about geoengineering and avoids a comprehensive assessment on the grounds of "limited evidence" (IPCC 2014a: 89). The

---

[9] A parallel study published by the NRC at the same time focused on carbon dioxide removal (CDR) geoengineering techniques.

document most read by those in power, the *Summary for Policymakers*, notes that "modelling indicates that SRM methods, if realizable, have the potential to substantially offset a global temperature rise" (IPCC 2013b: 29), whilst also pointing to a range of risks, including the high likelihood of significant termination effects.[10]

This is in contrast to the more obviously dismissive references to geoengineering in earlier IPCC reports. The Second Assessment Report, in its 'Summary for policymakers', concluded that geoengineering was "…likely to be ineffective, expensive to sustain and/or to have serious environmental and other effects that are in many cases poorly understood" (IPCC 1995c: 18). The Fourth Assessment Report, published shortly after the lifting of the taboo, makes no mention of geoengineering in its Synthesis Report (IPCC 2007a). It touches on ocean fertilization and solar radiation management in its Working Group 3 report. This noted that "[t]hese options tend to be speculative and many of their environmental side-effects have yet to be assessed; detailed cost estimates have not been published; and they are without a clear institutional framework for implementation" (IPCC 2007d: 624).

This ambivalence can also be found in the 2015 Paris agreement on climate change (UNFCCC 2015). The agreement makes no explicit mention of geoengineering. But it implicitly endorses, through its heavy reliance on 'negative emissions' and its assumptions regarding carbon-dioxide removal, the expanded take-up of CDR-type geoengineering. No mention is made of SGE. But by setting a temperature goal of 1.5C as the primary metric it is not surprising that some have seen implicit endorsement of SGE. As Richard Haass, president of the US Council on Foreign Relations (CFR) expressed it in a tweet following the release of the IPCC's 1.5 degree report: "Nothing suggests world will come close to meeting this goal on climate change. … We'd better set aside $ for adaptation and accelerate R&D on geoengineering" (Haass 2018). SGE can be understood to have a 'shadow presence' in the Paris agreement—it is both absent (and unspoken) and present (and inevitable?). It is a solution which dares not speak its name. I will return, in the concluding Chapter, to this 'geoengineering turn' in the Paris agreement.

---

[10] It should be noted that the outline for the next IPCC Assessment Report (AR6), due in 2021, suggests even greater attention will be given to geoengineering. It will be one of only eight cross-cutting issues to be covered in all three Working Group reports.

In Europe, most engagement with SGE has been in the UK and Germany. Papers were commissioned for the United Kingdom's parliament. But most influential was a 2009 report by the Royal Society (2009) entitled *Geoengineering the Climate: science, governance and uncertainty*. Without endorsing its deployment, this suggested that solar geoengineering could be both highly effective and highly affordable, arguing that it "may provide a potentially useful short-term backup to mitigation in case rapid reductions in global temperature are needed" (2009: 59). The UK also initiated the science council funded Integrated Assessment of Geoengineering Proposals (IAGP) in 2010, a Climate Geoengineering Governance project in 2012, and as well as the Stratospheric Particle Injection for Climate Engineering (SPICE) project. All have now concluded, the SPICE project being wrapped up prematurely in the face of much controversy and negative public reaction (Stilgoe 2015). The UK, which aspired to be a player in the SGE space is now largely an observer, although it continues to contribute to governance debates and some of its scientists remain connected to planned US experiments.

Germany has been the hub of European work on solar geoengineering, including a six-year inter-disciplinary priority programme, now extended, funded by the German Research Foundation—project SPP 1689. This is focused on assessment rather than deployment of geoengineering. It has concluded that SGE is less efficient, less effective and less safe than the Royal Society report suggested and that its side effects mean it should not be seen as a 'benign' intervention (Oschlies and Klepper 2017). In Germany, reports have been prepared for the Federal Ministry of Education and Research (Rickels et al. 2011) and the Umweltbundesamt (Ginzky et al. 2011), Germany's Environmental Protection Authority. Germany's policy focus, it would seem, is on watching and, where possible, shaping geoengineering's governance and restraining SGE's emergence. Elsewhere in Europe there has been a French programme Reflections on Environmental Geoengineering (REAGIR), and EU-funded programmes including The Implications and Risks of Engineering Solar Radiation to Limit Climate Change (IMPLICC) Project, and the European Transdisciplinary Assessment of Geoengineering (EuTRACE).

No major published 'official' or institutional studies are known to come from other globally powerful countries. In 2011, leading scientists from the SRMGI held high-level talks on SGE with Chinese scientists and officials (Edney and Symons 2014: 4). And in 2015 the Chinese government initiated a co-ordinated research programme on geoengineering (including solar geoengineering) as part of its National Key Basic Research Program. But the Chinese focus appears to be about engagement with geoengineering modelling globally and exploring the possible effects of SGE on key elements of the climate system, including precipitation and the monsoon (Cao et al. 2015; Moore et al. 2016). It is best seen as a watching brief. The Indian situation is similar. The government launched a research initiative in 2017 focused on the implications of SGE for developing countries, including possible effects on the global water cycle and extreme weather and cyclones, as well as on the governance of SGE (Bala and Gupta 2018).

In Russia, judging from the comments made by Russian 'official' bodies, such as the Institute for Global Climate and Ecology, regarding the drafts of the relevant chapters of IPCC's Fifth Assessment Report, enthusiasm for solar geoengineering runs high (IPCC 2013d, e). There have been attempts, including by Yuri Izrael, to conduct outdoor experiments spraying sulphate aerosols from helicopters and military trucks (Izrael et al. 2009).[11] But whilst enthusiasm is high in Russia, its capacity and influence are not.

In some middle-power countries, references to geoengineering are starting to appear in 'official' publications or policy documents aimed at the general public. An Australian example is a document 'The Science of Climate Change' published in February 2015 by the Australian Academy of Sciences, an updated version of a now established publication aimed at providing an accessible state-of-the-art account. On the final page, dealing

---

[11] Climatologist Yuri Izrael is a long-time enthusiast for solar geoengineering. A 2008 Russian report quotes him as saying "We must have different 'weapons' for fighting climate change and stabilizing the climate" and shows him to be a strong supporter of SGE as the "optimal and inexpensive" solution (Sinitsyna 2008). Izrael's position appears to be an example of the entwining of science and politics. He was a disciple of Budyko and co-authored *Global Climatic Catastrophes* with him (1988 [1986]). In the aftermath of the end of communism he was a one of the leaders of the Russian Academy and apparently close to Vladimir Putin. He was a vice-chair of the IPCC but in the early 2000s seems to have focused on discrediting any suggestion that climate change had anthropogenic causes and discouraging the Russian government from ratifying Kyoto (Bolin 2007: 187–9).

with options to address climate change, it states that managing climate risks "... will necessarily be based on some combination of four broad strategies". In addition to mitigation and adaptation, it also includes sequestration and solar geoengineering (AAS 2015: 30). This brings SGE into mainstream policy discussion. But middle-power countries are currently observers rather than 'players' in the solar geoengineering space.

In summary, there is not an embrace of geoengineering in 'official' circles, but the idea is being taken increasingly seriously. It has moved from taboo to seriously considered policy option. As aspiring solar geoengineer David Keith put it in his response to the NRC report: "it serves as a marker of the extent to which solar geoengineering is becoming a more normal part of the science and policy of climate change" (Keith 2015). A sign of the seriousness with which the idea is being taken can be found in its inclusion in new 'realist' climate policy thinking post-Copenhagen by individual academics known to influence policy shaping (for example Victor 2011; Wagner and Weitzman 2015) and post-Paris where some argue, not unreasonably, that SGE is "the only way to be certain of limiting global average temperature increases to 1.5°C" (MacMartin et al. 2018). Indeed it is here, as we shall see, in the 'policy influential' literature, less constrained by being the agreed product of committees of scientists and others, that a more consistent set of arguments about what geoengineering might be imagined to be, can appear.

It is notable that SGE is being taken most seriously in the US, which is no longer a party to the global climate agreement; but seen much more ambivalently by those committed to the global forums convened by the UNFCCC. Further, the depth of hostility towards SGE, even within the US, should not be underestimated. There too the idea of SGE, even as a policy option, faces significant resistance. As one climate scientist on the NRC's assessment report panel (2015), Raymond Pierrehumbert, put it:

> The nearly two years' worth of reading and animated discussions that went into this study have convinced me more than ever that the idea of "fixing" the climate by hacking the Earth's reflection of sunlight is wildly, utterly, howlingly barking mad. (Pierrehumbert 2015: n.p.)

3 The Re-emergence of Solar Geoengineering 95

To come to grips with what is at stake in geoengineering's re-emergence it is important to unpack how it is imagined in the 'official' and institutional literature. What are the expressed rationales for considering it? What narratives are used? What metaphors are mobilised? What visual representations predominate? How is it understood and framed? What unspoken assumptions underpin it? How solar geoengineering is imagined, especially given that it is a speculative technology, a policy proposal, will shape how, and even whether, it is undertaken. In particular, solar geoengineering assumes both that the dynamics of the climate system can be mastered sufficiently to engineer it, and that the relevant experts (scientists, engineers, institutions of governance, etc.) can be trusted to get it right. And yet narratives built on Mastery seem no longer viable, and trust in experts and elites (especially global ones) seems to be in decline.

Even as SGE has re-emerged the reservations of the Taboo era remain. Statements and propositions made are usually wrapped in cautious and conditional language. The most enthusiastic boosters of SGE often fail to convince fellow panellists when the institutional reports are compiled. Even in their own academic publications one observes that the dominant undercurrent is less enthusiastic embrace of geoengineering and more promotion of SGE's acceptance as inevitable necessity and 'realistic'. In another context, this has been called "conditional acceptance": it is reluctant, context-dependent, and subject to conditions (Bickerstaff et al. 2008).

Relatedly, the institutional reports almost always reveal fears about moving the climate policy focus away from emissions reduction, or even suggesting that mitigation is unlikely to work, although that is indeed what has prompted geoengineering's revival as an idea. The anxiety that geoengineering will divert attention from cutting emissions is commonly labelled as a 'moral hazard', which I examine in more detail in the next Chapter. For now it is sufficient to note most of the 'institutional' reports conclude with a *primary* recommendation prefacing their conclusions, of which the Royal Society report's 'Recommendation 1' is typical:

> Parties to the UNFCCC should make increased efforts towards mitigating and adapting to climate change and, in particular to agreeing to global emissions reductions of at least 50% of 1990 levels by 2050 and more thereafter. (2009: 57. See also NRC 2015: 3)

We can now turn to a closer examination of the rationales, frames and representations which have accompanied SGE's re-emergence. Our focus is on the 'official' and institutional literature. More oppositional accounts will be explored in the next Chapter.

Rationales

Since geoengineering's re-emergence three basic rationales for SGE can be identified. These are: climate emergency, climate risk reduction, and alternate climate Plan A (alternate to emissions reduction that is). Table 3.1 provides a summary outline of these. It should be noted that in 'official' studies the rationales are rarely fleshed out, are often used in combination, and occasionally unstated or only implicit.

Climate Emergency

The climate emergency rationale sees solar geoengineering, as a tool to be used in the case of a *specific* climate emergency, to hold off or reverse 'tipping' into rapid warming and/or imminent climate catastrophe. The Royal Society report predominantly takes this approach: SGE "may provide a potentially useful short-term backup to mitigation in case rapid reductions in global temperature are needed" (2009: 59). The logic here is that the technology should be researched and developed and then kept in the toolbox for use if it is, regrettably, needed. The 2009 Novim report, entitled 'Climate Engineering Responses to Climate Emergencies', and produced by an eminent panel of policy experts and scientists, even includes "emergency" in the report title (Blackstock et al. 2009).

As we shall see in Chapter 5, this rationale struggles to withstand critical interrogation. What constitutes an emergency? Who proclaims it? Has it already started? Can a climate emergency be predicted? ... and so on. Nevertheless, climate emergency remains a common rationale, especially in presentation of the subject by and to non-specialists, and in the popular imagination. It is also the rationale used by the Arctic Methane Emergency Group (AMEG), a grouping comprised mainly of concerned polar scientists.

3 The Re-emergence of Solar Geoengineering   97

Table 3.1 Typology of arguments in favour of solar geoengineering

	Climate emergency (Plan B)	Risk reduction (Extra policy tool)	Alternative climate plan (Better policy)
Reasoning	May need it to hold off imminent catastrophe or reverse 'tipping' into rapid warming	Slow warming and buy time (to cut emissions, develop new technologies & overcome policy inertia)	Cheaper and more socially, economically and politically acceptable than abatement
Implications for abatement	Abatement essential but may not happen in time	Abatement essential but easier later. GE may supplement modest abatement targets	GE could be a substitute for much abatement
When deployed?	When necessary and for as long as emergency lasts	As soon as ready and for as long as needed	As soon as ready and then indefinitely
In practice	Support research now to ensure future readiness	Research then deploy incrementally	Research then deploy
Typical metaphors	Insurance policy, parachute	Chemotherapy, planetary medicine	Sunscreen, innovation
Attracts	Climate catastrophists & technophiles	Climate policy 'realists' & ecomodernists	Market fundamentalists and climate inactivists

They have argued that "the tipping point for the Arctic sea ice has already passed" and that this is "a catastrophic threat for civilisation" (AMEG 2014: n.p.). AMEG accuses the IPCC of, in its Fifth Assessment Report, vastly underestimating the rising concentrations of methane, and argues that "[i]mmediate action must be taken to refreeze the Arctic to halt runaway melting" (AMEG 2014: n.p.). In short, they favour solar geoengineering now.[12] In his pivotal paper lifting the geoengineering taboo,

---

[12] AMEG's website reveals the mood of despair amongst leading Arctic scientists. Undoubtedly heartfelt, the AMEG approach is socially and politically naïve, in the way it imagines a science-led climate policy.

Crutzen also relied on the emergency rationale: "to combat potentially drastic climate heating" (2006: 216). But it is less commonly used in the more recent 'institutional' literature.

Risk Reduction

A second rationale focuses on geoengineering as a form of climate risk reduction, to shave the peaks off the warming effect or slow the rate of warming, thereby both supplementing and allowing time for mitigation through emissions reduction and adaptation. The approach of the NRC study tends towards the risk argument. It does not adopt a detailed rationale but recognises that mitigation has been limited to date, and that "… it may be prudent to examine *additional* options for limiting the *risks* from climate change… [as part of] a broader portfolio of responses" (2015: 2, my emphasis). The GAO report (2011) argues along similar lines. So too does the IPCC: in 2010 it expressed the need for an expert group to investigate "… possible geoengineering options to *complement* climate change mitigation efforts" (see IPCC 2012: 10, my emphasis).

The logic of the risk reduction rationale is to research and develop the technology and, if sufficiently safe, use it *now* to reduce/constrain warming, and do so as part of a combination of actions including emissions reduction. This was the approach suggested by leading climate scientist Tom Wigley in his 2006 intervention in the NASA-Ames symposium discussed above. In another publication that same year, entitled 'A Combined Mitigation/ Geoengineering Approach to Climate Stabilization', Wigley argued that "[m]itigation is therefore necessary, but geoengineering could provide additional time to address the economic and technological challenges faced by a mitigation-only approach" (Wigley 2006: 452). Similar arguments are at the heart of David Keith's *A Case for Climate Engineering* (2013).

Alternate Climate Policy

The third rationale sees geoengineering as an alternative climate plan, attractive because it is cheaper, less economically disruptive and therefore more politically acceptable than the current Plan A (mitigation). In this view solar geoengineering could largely substitute for abatement in the short to medium term since its benefits exceed its costs. The logic is that the technology should be developed and deployed and if emissions reductions are still needed then this can be addressed later when new technology will have made abatement cheaper. The alternate Plan A rationale is rarely found explicitly in the major 'institutional' reports, although when these emphasise SGE's cheapness they implicitly adopt this line of thinking. A rare example in the 'institutional' literature can be found in the perspectives of Lowell Wood voiced at the Tyndall-Cambridge-MIT symposium, presented and discussed above.

More commonly this rationale is voiced by free-market think-tanks and advocacy groups such as in the report of the American Enterprise Institute (Lane and Bickel 2013) and the related Copenhagen Consensus Center report on geoengineering which claims to have made "… a strong case that the potential net benefits of SRM are large" (Bickel and Lane 2009: 52); or when conservative US politician Newt Gingrich, notorious for flip-flopping on climate change, says that "geoengineering holds forth the promise of addressing global warming concerns for just a few billion dollars a year" (2008; see also Jackson 2011). The 'alternative climate plan' rationale differs from the previous two rationales identified in that many of its proponents regard as exaggerated the claim that climate change is a serious problem.

These rationales are often used in combination. For example, the Royal Society report supplements its central (emergency) argument when, primarily in relation to CDR techniques rather than SGE, it notes the merits of "… a portfolio approach to climate change, [where] properly researched geoengineering methods …. could eventually be useful to augment conventional mitigation activities, even in the absence of an imminent emergency" (Royal Society 2009: 56). Even Lane and Bickel's more recent work for the American Enterprise Institute, with its focus on the "… political bankruptcy of GHG control policies [and] SRM's economic promise" (2013: 20), repositions its support for solar geoengineering as potentially "… a highly useful backup and supplement to current policy options" (p. iv). The Bipartisan Policy Center believes that managing risk is the central principle of effective climate policy. It argues that 'climate remediation', its preferred term for geoengineering, is appropriate only as a complementary measure or in the event of an emergency (BPC 2011: 3). And the IPCC's Fifth Assessment Report,

Working Group 3, frames geoengineering mainly as a risk issue but nevertheless manages to retain emergency argumentation when it argues that "… [risk] strategies require preparing for possible extreme climate risks that may implicate the use of geoengineering technologies as a last resort in response to climate emergencies" (IPCC 2014c: 114). The addition of the standard IPCC formulation "limited evidence, low agreement" to this statement illustrates the still contested nature of the geoengineering project.

Frames and Metaphors

Frames, metaphors and their accompanying narratives shape how we see and imagine things and how we might act. They can have powerful effects, particularly in the case of a technology which is not yet operational, and where its emergence is contested. Metaphors are rhetorical devices, presenting one thing in terms of another, and thereby providing "visions of the world and instruments to change it" (Nerlich and Jaspal 2012: 133). To frame, in Entman's classic formulation, "… is to select some aspects of a perceived reality and make [it] more salient … in such a way as to promote a particular problem definition, causal interpretation, moral evaluation, and/or treatment recommendation for the item described" (1993: 52).

Geoengineering is, of course, part of a larger discourse about climate change. Four elements from that larger discourse are relevant here. Firstly, one of the key narratives which emerges in international climate policy from the early 2000s is of 'dangerous climate change', as something to be avoided, and indeed to be fought. Shaw and Nerlich show that, whilst there are earlier antecedents, in the 2003–07 period themes and metaphors of 'dangerous limits', 'ticking clocks', 'thresholds' and 'combating', 'attacking' and 'fighting' climate change come to the fore (2015: 37). This was the period of SGE's re-emergence and reinforced its emergency rationale both in general and in particular. In the scientific literature, important visual representations which reinforced the sense of imminent danger were the 'burning embers' image, common since it first appeared in the IPCC's 3rd Assessment Report in 2001; and the 'tipping point' cartography which emerged from around 2004 (Kemp 2005).

Secondly, as Liverman points out, "… these powerful images … are predominantly biophysical, with human systems and geographies relatively unexplored or obscured" (2009: 287–8). The implication is that 'climate emergency', whilst it may be anthropogenically driven, remains primarily an object of science rather than a socially, temporally and spatially located entity. As Hulme has argued, since the late 1980s the dominant framing of climate change has been as a largely physical phenomenon (2008; see also Hulme 2009). This framing sits easily with both geoengineering's emergency and risk reduction rationales. And, as we shall see below, this is manifest in the ways in which geoengineering is visually represented in the 'official' literature.

Thirdly, there have been efforts to make the emissions reduction idea compatible with neo-liberal market discourse with its emphasis on economic growth. The idea that 'stemming the rise in emissions need not cost the Earth' was a central trope of the 2007 IPCC Fourth Assessment Reports. Indeed Shaw and Nerlich have identified a shift in the 'official' climate narrative between 2005 and 2007 which "… reframed the climate mitigation discourse in line with the demands of GDP growth" (2015: 35). These framing turns sit most comfortably with both geoengineering's 'risk reduction' and its 'alternative climate plan' rationales.

Finally, an influential role has been given to science and scientists in shaping climate policy. Figure 3.1, a cartoon illustration from *Nature* captures the metaphorical zeitgeist. It depicts three scientists—Bert Bolin (meteorologist and chair of the IPCC until 1997), John Shepherd (oceanographer and then deputy director of the Tyndall Centre and leader of the Royal Society investigation into geoengineering), and Luiz Gylvan Meira Filhov (climatologist and former co-chair of the science working group of the IPCC)—diagnosing Earth's carbon-induced fever.[13] Metaphorically and visually the planet is presented as a critically ill patient, and the scientists as physicians responsible for monitoring, diagnosis and cure.

There is now a substantial literature examining metaphors and framing in relation to geoengineering, and solar geoengineering in particular (for example Cairns and Stirling 2014; Corner and Pidgeon 2015; Macnaghten and Szerszynski 2013; Nerlich and Jaspal 2012; Shaw and Nerlich 2015; Markusson et al. 2014; Markusson 2013; Porter and Hulme 2013; Sikka

---

[13] I acknowledge Jim Fleming for drawing this illustration to my attention.

Fig. 3.1 Planet as patient, scientist as physician (Artwork with artist's permission: David Parkins, from *Nature*, (2008) *455*(7214), 737.)

2012; Huttunen and Hilden 2014). A study by Nerlich and Jaspal of the use of argument and metaphor related to geoengineering in the period up to 2010

> revealed one master argument (The earth is seriously/catastrophically/broken/ill and can only be fixed/healed by geoengineering) which was linked to three conceptual master metaphors: "THE PLANET IS A BODY," "THE PLANET IS A MACHINE," "THE PLANET IS A PATIENT." The Persuasive force of this discourse emerges from a fusion of the master-argument with the master-metaphors. It exploits what Beck called "the political potential of catastrophes". (2012: 141)

Their assessment is borne out by a close reading of the institutional literature, all of which stresses the dangerous climate situation and the risks this poses. In the foreword to The Royal Society report by its President, Lord Rees, we are warned that if reductions "achieve too little, too late" then a "Plan B" may be needed; although he stresses the need to be aware that no geoengineering option offers "a silver bullet" (2009 :v). The technology assessment conducted by the US GAO speaks of "an insurance policy" against "worst case climate scenarios" (2011: ii).

Similarly, the Bipartisan Policy Center makes use of the language of risk and of "tipping points" (2011: 3–4). Its insistence on calling geoengineering "climate remediation" is itself suggestive of a vision which sees the climate as damaged but fixable, and able to be returned through geoengineering to something approximating its former state.

But, these examples apart, the 'institutional' literature, with few exceptions, does not utilise metaphors extensively. To get a deeper insight we need to draw on the metaphors used by leading scientists central to the development and drafting of these reports. Crutzen, in his 2006 article, argued the need for "an escape route" (2006: 216). "If my plane is going down in flames", says Ken Caldeira, "I sure hope I have a parachute handy" (Caldeira cited in David 2007: 32). Medical metaphors and emergency framings are especially commonplace. "If you have a heroin addict", argued Stephen Schneider, "the correct treatment is hospitalisation, therapy and a long rehab. But if they absolutely refuse, methadone is better than heroin" (quoted in Chandler 2007: 43). It is "like chemotherapy", David Keith is reported as saying. "No one wants to have it ... but we all want the ability to do chemotherapy and know its risks should we find ourselves with cancer" (cited in Howell 2010). The President of the Royal Society, writing in support of a geoengineering experiment (the SPICE project), argued that "[g]eoengineering research can be considered analogous to pharmaceutical research" (Nurse 2011). James Lovelock refers to geoengineering as "planetary medicine" (2008, 2009). A related, persistent metaphor is of solar geoengineering as a "sunscreen" or "sunshade" (Teller 1997; Chandler 2007). Caldeira again: "If we become addicted to a planetary sunshade, we could experience a painful withdrawal if our fix was suddenly cut off" (cited in David 2007: 32). As Nerlich and Jaspal note, there is a tendency to "… supplement the argument from catastrophe with metaphors and analogies of healing and medicine" (2012: 143).

Less commonly, machine metaphors are found, such as conceiving of geoengineering as an "emergency brake" (Brovkin et al. 2009), or referring, in this instance negatively, to "retooling the planet" (Bronson et al. 2009). Significantly, the influential NRC report asks the reader to imagine the Earth as a machine, specifically a thermostat.

> The climate system can be compared to a heating system with two knobs, either of which can be used to set the global mean temperature. The first knob is the concentration of greenhouse gases ... The other knob is the

reflectance of the planet, which controls the amount of sunlight that the Earth absorbs.... [T]hese two knobs do more than affect global mean temperature. In differing ways, they also influence regional temperatures, the global hydrological cycle, land plants, and other components of the Earth system. So, turning up one knob and turning down the other might be able to restore Earth's global mean temperature, but could nevertheless produce substantial changes to Earth's environment. (2015: 27–8)

Scientists involved in the institutional reports who are less favourably disposed to SGE invoke their own metaphors. These sometimes draw on the same master metaphors—"a band-aid to buy time", "not a silver bullet", "not a single global thermostat"—but also use phrases like "playing god", "geopiracy" or "techno-fix" (Nerlich and Jaspal 2012: 141–2). Interviewed in 2007, Meinrat Andreae, an atmospheric scientist at the Max Planck Institute for Chemistry viewed geoengineering as way to feed our addiction to fossil fuels: "[i]t's like a junkie figuring out new ways of stealing from his children" (cited in Morton 2007). Here SGE is the disease not the cure.

In June 2006, Gavin Schmidt, a climate modeller at NASA, wrote a skeptical blog post in anticipation of the publication of Crutzen's taboo-breaking paper in *Climatic Change*. The analogy he uses is worth citing at length, as it captures much of the negative attitude towards geoengineering then (and still?) prevailing among many climate scientists:

Think of the climate as a small boat on a rather choppy ocean. Under normal circumstances the boat will rock to and fro, and there is a finite risk that the boat could be overturned by a rogue wave. But now one of the passengers has decided to stand up and is deliberately rocking the boat ever more violently. Someone suggests that this is likely to increase the chances of the boat capsizing. Another passenger then proposes that with his knowledge of chaotic dynamics he can counterbalance the first passenger and indeed, counter the natural rocking caused by the waves. But to do so he needs a huge array of sensors and enormous computational resources to be ready to react efficiently but still wouldn't be able to guarantee absolute stability, and indeed, since the system is untested it might make things worse. So is the answer to a known and increasing human influence on climate an ever more elaborate system to control the climate? Or should the person rocking the boat just sit down? (Schmidt 2006: n.p.)

Representations

Visual representations of geoengineering in the 'institutional' literature are remarkably uniform across the reports. Figure 3.2 comes from the Lawrence Livermore National Laboratory (LLNL) and is widely reproduced in assessment reports. Figure 3.3 was used to illustrate the IPCC's Fifth Assessment Report section on geoengineering (2013a: 632). Both are typical examples. All present an un-peopled world, and the engineering of the climate is portrayed as a technical, physical endeavour with a range of interventions possible. In all the images geoengineering is primarily about particular categories of technological intervention, and these are conceived as technoscientific objects (see also Schrickel 2014).

Figure 3.4 shows how SGE is represented in the NRC report (2015: 33). Similar representations exist in all the reports.[14] The emphasis is entirely physics-focussed. Solar radiation management is visualised as

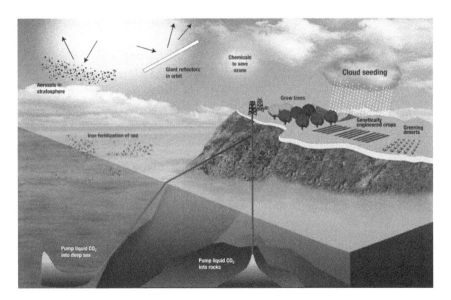

Fig. 3.2 A standard depiction of geoengineering (Source: Lawrence Livermore National Laboratory)

---

[14] See for example Royal Society (2009: 23); Umweltbundesamt (Ginzky et al. 2011: 11); and Bipartisan Policy Center (2011: 25).

106 J. Baskin

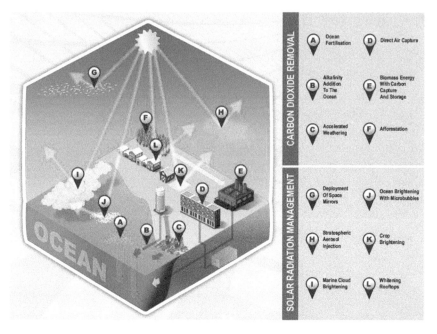

Fig. 3.3 Representation of geoengineering in IPCC 5th Assessment Report (2013a: 632)

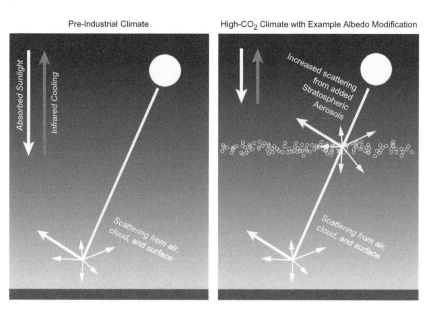

Fig. 3.4 Visualisations of SGE in NRC Report (2015: 33)

3 The Re-emergence of Solar Geoengineering

little more than a technical option for intervention in the radiation balance of the Earth, changing the existing formula regarding how much radiation comes in and goes out.

An engineering focus is evident when it comes to depicting the delivery of aerosols into the stratosphere. Figure 3.5 presents one such envisioning, in this case drawn from the SCoPEx project. Beyond this experiment, the use of a fleet of aeroplanes to deliver the aerosols is the currently favoured approach. Other envisionings in the 'policy-influential' literature are along similar lines.[15]

All these representations reinforce the notion of climate change as primarily a physical object, with the corollary that SGE is primarily a scientific concern and an engineering praxis. Absent from these images is the suggestion of any negative (or even positive) effects on actual people.

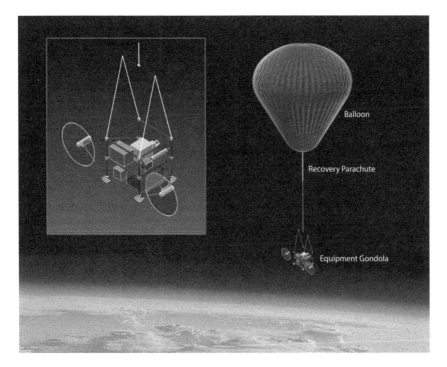

Fig. 3.5   Image from SCoPEx project website (Courtesy: Harvard University)

---

[15] See for example Robock et al. (2009) which included guns, towers, balloons and aircraft as possible delivery mechanisms, and the illustration of the SPICE project in Macnaghten and Owen (2011).

Indeed there are no people in these images at all, and no social systems are represented or imagined. Absent too is anything comparable to the 'heroic' scientist of the Mastery phase depicted in Chapter 2 (Fig. 2.3). Indeed the viewer watches from afar or hovers, unseen but all-seeing, god-like, over the world.

An Inability to Normalise

Since the mid-2000s geoengineering has moved into mainstream climate policy consideration but SGE, as a proposition, has struggled to become normalised as a part of the climate policy landscape. It remains highly contested. One symptom of this can be seen in the inability even to settle on an agreed name for SGE. Another can be found in the post-Paris representation of negative emission technologies as "not geoengineering", in an effort to distance these from SGE, despite having been widely categorised as geoengineering previously.

Finnemore and Sikkink (1998) have described and theorised the emergence of norms, and their ratification in international treaties. The first of the three stages they identify, 'norm emergence', envisages persuasion by 'norm entrepreneurs' as central. Using this typology, SGE has clearly already emerged, in that it is under active consideration and is no longer 'taboo'. But it has not reached a 'tipping point', as they describe it, leading to stage two, where it disperses and 'cascades' outwards as a norm. Indeed, those pushing for SGE have faced significant pushback. SGE is some distance, therefore, from becoming 'internalized' and essentially taken for granted or regarded as self-evident (stage three). By contrast, using Finnemore and Sikkink's schema, climate mitigation through emissions reduction could be regarded as a widely-dispersed norm, and substantially taken-for-granted (if not always acted upon).

As Finnemore and Sikkink point out, "[m]any emergent norms fail to reach a tipping point" (1998: 895). SGE has manifestly not yet crossed the norm tipping point. Certainly, geoengineering has its enthusiasts and proponents, its 'norm entrepreneurs': the knowledge-brokers described in the Introductory Chapter. But they are not, currently, persuading a sufficiently large or important enough constituency for norm 'cascading' to have occurred.

The disagreement about SGE among climate scientists and climate policymakers can be seen, for example, in the comprehensive NRC study which decided to treat 'albedo enhancement' as distinct from other forms of geoengineering (even isolating it into a separate report). The report's summary "… reiterates that it is opposed to climate-altering deployment of albedo modification techniques…" (p. 11), but elsewhere it qualifies this by recommending that solar geoengineering "should not be deployed at this time" (2015: 7). It recommends further research, but this is hedged by the need for this to be "multiple benefit research" which enhances climate science generally (p. 9), rather than specifically focussed on solar geoengineering. It also argues that any research governance should ensure "civil society involvement" (p. 10), and that what is researched should be the subject of "a serious deliberative process" (p. 10). On one reading, this is simply institutional language in which compromise wording is used to displace controversy, and where disagreements are smoothed over and postponed in the call for further research. But it can also be interpreted as coming remarkably close to a position of saying there should be no further SGE research (Pierrehumbert 2015).

Whilst there are no published studies to confirm this, my own observations and discussions with others who are knowledgeable, suggest that only a small minority of climate scientists favour SGE, and that this remains a minority even if one adds those who express a favourable attitude to SGE conditional upon their scientific and governance concerns about it being satisfactorily addressed. Many scientists cite concerns about the proposed technology's efficacy and conclude that solar geoengineering is not a wise course of action—that the technology's unreliability is a major cause for concern (Robock 2008). Others voice concern about what SGE would lock in and the governance implications: "it is not possible to use albedo modification to counteract peak CO_2-induced warming without maintaining the climate intervention without interruption for millennia" (Pierrehumbert 2015). Alan Robock, a leading climate scientist and volcano expert, has listed '20 reasons why geoengineering may be a bad idea' (2008) and more recently has come up with 5 benefits and 26 risks (2014: n.p.). His reasons are largely practical and range from SGE's workability to its (non-)reversibility, from unknown impacts to more niche concerns such as the "degradation of terrestrial optical astronomy". But Robock's list of objections also includes "Moral authority: Do we have the right to do this?" (2014: n.p.). The charge of 'hubris' implicit in this objection is frequently made. It underpins Hulme's study where he argues forcefully that solar geoengineering is "undesirable, ungovernable and unreliable" (2014), and lies at the heart of much public hostility and NGO opposition to SGE.

In some respects, the inability to normalise SGE to date is surprising. After all, it has re-emerged out of the existing domains of expertise on climate change. It is compatible with some of the dominant narratives and framings of climate change already described: of impending climate crisis requiring 'action', of climate being understood primarily as a physical phenomenon (notwithstanding the anthropogenic attribution), and of the compatibility of the economic growth and the climate action agendas. And it seems to address real concerns that the main policy response, mitigation, is not delivering declines in or even the stabilisation of GHG concentrations.

How can the failure of SGE to become normalised be explained? One can identify four reasons. These, and others, will be explored further in later Chapters.

Firstly, values-contestation is at the heart of contestation over SGE. The rationales for embarking on solar geoengineering outlined earlier rest on contested values and assumptions: assumptions about emergency and risk, economy and development, 'nature' and society, about what ethical/values questions are most important, and for whose benefit and at whose behest might SGE be legitimately deployed. SGE aims to remake the climate and this brings to the fore differing conceptions of how the world ought to be. Such considerations are either ignored (as being outside the brief) or displaced (acknowledged and put aside) when the major institutional reports present SGE as a mundane technology, a technoscientific object. For example, the Royal Society report acknowledges hubris, but in its brief consideration of ethics, it categorises such concerns as part of "virtue ethics", and seemingly prefers to focus on other philosophical traditions "… where a consequentialist case in favour can be made" (2009: 39). The GAO report notes that "a small number of the experts we consulted" viewed climate engineering as "technological hubris" and, here the tone of the report appears shocked, that some even "opposed starting significant research" (2011: 53). Whilst some proponents reject or sidestep the charge of hubris (see, for example, Keith et al. 2010: 426), others share such anxieties (for example Schneider 2001). Substantial value differences are not easily resolved.

Secondly, for an idea to be normalised it needs to be presented as something desirable. The idea of geoengineering is difficult to frame in this way, at least when presented in isolation from other positive visions of the future. Solar geoengineering is generally presented as 'sadly inevitable', as a pessimistic technology: a dystopian alternative to a dystopian climate future. In later Chapters I will explore some exceptions, which try to attach SGE to developmentalist and 'good Anthropocene' discourses. But such arguments have, to date, been rare and the overall tone in discussions of SGE is gloomy.

Thirdly, for geoengineering to become normalised means diluting (at the very least) one of the key existing norms of climate policy: the need to mitigate by cutting emissions. The 'alternative climate plan' rationale outlined above explicitly presents SGE as a possible alternative to the focus on emissions reduction. But the other two rationales also imply some dilution of mitigation. By adding a new tool to climate strategy, they thereby make mitigation simply one of a number of policy actions. The repeated assertions of 'Recommendation 1' in the 'official' assessments— that nothing said in the 'institutional' reports should be interpreted in a way that undermines mitigation, and that SGE, if deployed, should be seen as a supplement to mitigation—is an attempt to allay precisely such concerns. But the assertion itself does not make it 'true'. Indeed, the heatedness of debates around whether geoengineering would discourage (or enhance) mitigation can be seen as recognition that solar geoengineering, whether intended or not, brings mitigation's centrality into question. A great deal is at stake in seemingly arcane discussions of 'moral hazard'.

According to Finnemore and Sikkink (1998), the dominant mechanisms through which norms 'cascade' are socialisation, institutionalisation and demonstration. Being officially regarded as a third leg of climate policy would go a long way towards the institutionalisation and socialisation of SGE. But this recognition is still being withheld by the key institutions, such as the IPCC. Which brings us to the fourth reason: solar geoengineering as technology lacks a clear demonstration path. It is still a technology operating in the space between imagination and demonstration. It is more than a speculative technology, in that time, money, research and limited trials are involved too. But it does not yet exist as an operating, deployable technology. It can be imagined but it is not easily assessed, and assessment to date relies heavily on modelling. This requires a high level of confidence in climate models which themselves are prisoners of their own assumptions, as I shall explore in the next Chapter. Indeed, arguably, SGE can only be assessed through practical experiment, using trial and error methods and making adjustments along the way. In the absence of field experiments (which key proponents are urging), Finnemore and Sikkink's demonstration mechanism is effectively unavailable.

All these reasons have contributed to the inability of SGE to become normalised and an inability to push dissenting voices to one side. SGE has now entered the climate policy mainstream. It has brought to the surface a range of knowledge differences regarding the climate prognosis and what SGE's effects are likely to be; value differences about what is and is not acceptable for humans (and scientists) to do; and power differences regarding whose interests are furthered by SGE and whose are harmed. These differences will not evaporate if only more research could be done, and 'knowledge gaps' filled. Clearly bigger issues are at play than disputes about the efficacy of the proposed technology itself. How these are resolved will depend heavily on which narratives and imaginaries take hold. It is to the competing imaginaries that we now turn.

References

AAS (Australian Academy of Science). (2015). *The science of climate change*. Canberra: Australian Academy of Science.
AMEG (Arctic Methane Emergency Group). (2014). Press release, 4 December. Retrieved September 20, 2016, from http://ameg.me/
Bala, G., & Gupta, A. (2018). India's push for solar geoengineering. *Nature, 557*, 637.
Belter, C. W., & Seidel, D. J. (2013). A bibliometric analysis of climate engineering research. *WIREs Climate Change, 4*(5), 417–427.

Bernstein, S., & Ivanova, M. (2008). Institutional fragmentation and normative compromise in global environmental governance: What prospects for re-embedding? In S. Bernstein & L. W. Pauly (Eds.), *Global liberalism and political order: Toward a new grand compromise?* New York: State University of New York Press.

Bickel, J. E., & Lane, L. (2009). *An analysis of climate engineering as a response to climate change.* Copenhagen: Copenhagen Consensus Center.

Bickerstaff, K., Lorenzoni, I., Pidgeon, N., Poortinga, W., & Simmons, P. (2008). Reframing nuclear power in the UK energy debate: Nuclear power, climate change mitigation and radioactive waste. *Public Understanding of Science, 17*, 145–169.

Bipartisan Policy Center (BPC). (2011). *Geoengineering: A national strategic plan for research on the potential effectiveness, feasibility, and consequences of climate remediation technologies.* Washington, DC: Bipartisan Policy Center Task Force on Climate Remediation Research.

Blackstock, J. J., Battisti, D. S., Caldeira, K., Eardley, D. M., Katz, J. I., Keith, D. W., et al. (2009). *Climate engineering responses to climate emergencies.* Santa Barbara: Novim. Retrieved January 9, 2019, from http://arxiv.org/pdf/0907.5140

Bolin, B. (2007). *A history of the science and politics of climate change.* Cambridge: Cambridge University Press.

Bowen, M. (2008). *Censoring science: Inside the political attack on Dr. James Hansen and the truth of global warming.* New York: Dutton.

BPC. (2011). See Bipartisan Policy Center 2011.

Branson, R. (2008). Addressing climate change: Keynote address by Sir Richard Branson. Retrieved January 9, 2019, from http://www.un.org/ga/president/62/ThematicDebates/statements/RichardBransonSpeech.shtml

Bronson, D., Mooney, P., & Wetter, K. J. (2009). *Retooling the planet? Climate chaos in the geoengineering age.* A report prepared by ETC Group for the Swedish Society for Nature Conservation. Stockholm: Swedish Society for Nature Conservation. Retrieved January 9, 2019, from http://www.etcgroup.org/content/retooling-planet-new-etc-group-report-geoengineering

Brovkin, V., Petoukhov, V., Claussen, M., Bauer, E., Archer, D., & Jaeger, C. (2009). Geoengineering climate by stratospheric sulfur injections: Earth system vulnerability to technological failure. *Climatic Change, 92*(3–4), 243–259.

Brysse, K., Oreskes, N., O'Reilly, J., & Oppenheimer, M. (2013). Climate change prediction: Erring on the side of least drama? *Global Environmental Change, 23*, 327–337.

Budyko, M. I., Golitsyn, G. S., & Izrael, Y. A. (1988 [1986]). *Global climatic catastrophes*. Translated from Russian by V. G. Yanuta. New York: Springer Verlag.

Cairns, R., & Stirling, A. (2014). 'Maintaining planetary systems' or 'concentrating global power?' High stakes in contending framings of climate geoengineering. *Global Environmental Change, 28*, 25–38.

Cao, L., Gao, C.-C., & Zhao, L.-Y. (2015). Geoengineering: Basic science and ongoing research efforts in China. *Advances in Climate Change Research, 6*(3–4), 188–196.

Chandler, D. L. (2007, July 21). A sunshade for the planet: If we can't stop global warming, as a last resort, researchers are devising ways to cool the planet by shading it from the sun. *New Scientist*, pp. 42–45.

Christoff, P., & Eckersley, R. (2013). *Globalization & the environment*. Lanham, MD: Rowman and Littlefield.

Cicerone, R. J. (2006). Geoengineering: Encouraging research and overseeing implementation. *Climatic Change, 77*, 221–226.

Corner, A., & Pidgeon, N. (2015). Like artificial trees? The effect of framing by natural analogy on public perceptions of geoengineering. *Climatic Change, 130*(3), 425–438.

Crist, E. (2007). Beyond the climate crisis: A critique of climate change discourse. *Telos, 141*(Winter), 29–55.

Crutzen, P. J. (2006). Albedo enhancement by stratospheric sulfur injections: A contribution to resolve a policy dilemma? *Climatic Change, 77*(3–4), 211–219.

Dalby, S. (2009). *Security and environmental change*. Cambridge: Polity.

David, L. (2007, September). Climate change: A geoengineering fix? *Aerospace America, 45*, 32–37.

Dibley, B., & Neilson, B. (2010). Climate crisis and the actuarial imaginary: The war on global warming. *New Formations, 69*(Spring), 144–159.

Dickinson, E. (2010, September 3). A changed climate skeptic? *Foreign Policy*.

Edney, K., & Symons, J. (2014). China and the blunt temptations of geoengineering: The role of solar radiation management in China's strategic response to climate change. *The Pacific Review, 27*(3), 307–332.

Entman, R. (1993). Framing: Toward clarification of a fractured paradigm. *Journal of Communication, 43*(4), 51–58.

Finnemore, M., & Sikkink, K. (1998). International norm dynamics and political change. *International Organization* (Special issue: 'International Organization at Fifty: Exploration and Contestation in the Study of World Politics'), *52*(4), 887–917.

GAO (Government Accountability Office). (2010). *Climate change: A coordinated strategy could focus federal geoengineering research and inform governance efforts.* GAO-10-903. Washington, DC: U.S. Government Accountability Office. Retrieved January 9, 2019, from https://www.gao.gov/assets/320/310105.pdf

GAO (Government Accountability Office). (2011). *Climate engineering: Technical status, future directions, and potential responses.* GAO-11-71. Washington, DC: U.S. Government Accountability Office. Retrieved January 9, 2019, from http://www.gao.gov/new.items/d1171.pdf

Gingrich, N. (2008, June 3). Stop the green pig: Defeat the Boxer-Warner-Lieberman green pork bill capping American jobs and trading America's future. *Human Events.* Retrieved January 9, 2019, from http://humanevents.com/2008/06/03/stop-the-green-pig-defeat-the-boxerwarnerlieberman-green-pork-bill-capping-american-jobs-and-trading-americas-future/

Ginzky, H., Herrmann, F., Kartschall, K., Leujak, W., Lipsius, K., Mäder, C., et al. (2011). *Geoengineering: Effective climate protection or megalomania?* Dessau-Roßlau: Umweltbundesamt.

Gore, A. (2004). The climate emergency. In J. R. Lyons, H. S. Kaplan, F. Strebeigh, & K. E. Campbell (Eds.), *Red, white, blue, and green: Politics and the environment in the 2004 election.* New Haven, CT: Yale School of Forestry & Environmental Studies.

Haass, R. (2018, October 7). Retrieved January 9, 2019, from https://twitter.com/RichardHaass/status/1049107983618252800

Howell, K. (2010, February 5). Climate: Scientists call for interagency geoengineering research program. *Environment and Energy Daily.*

Hulme, M. (2008). Geographical work at the boundaries of climate change. *Transactions of the Institute of British Geographers, 33*(1), 5–11.

Hulme, M. (2009). *Why we disagree about climate change: Understanding controversy, inaction and opportunity.* Cambridge: Cambridge University Press.

Hulme, M. (2014). *Can science fix climate change? A case against climate engineering.* Cambridge: Polity Press.

Huttunen, S., & Hilden, M. (2014). Framing the controversial: Geoengineering in academic literature. *Science Communication, 36*(1), 3–29.

IPCC (Intergovernmental Panel on Climate Change). (1995a). *Second Assessment Climate Change 1995, synthesis report.* Geneva: IPCC.

IPCC (Intergovernmental Panel on Climate Change). (1995c). *Second Assessment Report, summary for policymakers.* Geneva: IPCC.

IPCC (Intergovernmental Panel on Climate Change). (2001). Climate Change 2001: Synthesis report. In R. T. Watson & The Core Writing Team (Eds.), *A contribution of Working Groups I, II, and III to the Third Assessment Report of the Intergovernmental Panel on Climate Change*. Cambridge, UK: Cambridge University Press.

IPCC (Intergovernmental Panel on Climate Change). (2007a). Climate Change 2007: Synthesis report. In R. K. Pachauri & A. Reisinger (Eds.), *Contribution of Working Groups I, II, and III to the Fourth Assessment Report of the Intergovernmental Panel on Climate Change*. Geneva: IPCC.

IPCC (Intergovernmental Panel on Climate Change). (2007d). Climate Change 2007: Mitigation. In B. Metz, O. R. Davidson, P. R. Bosch, R. Dave, & L. A. Meyer (Eds.), *Contribution of Working Group III to the Fourth Assessment Report of the Intergovernmental Panel on Climate Change*. Cambridge, UK and New York: Cambridge University Press.

IPCC (Intergovernmental Panel on Climate Change). (2012). *Meeting report of the Intergovernmental Panel on Climate Change expert meeting on geoengineering* (O. Edenhofer, R. Pichs-Madruga, Y. Sokona, C. Field, V. Barros, T. F. Stocker, Q. Dahe, J. Minx, K. Mach, G.-K. Plattner, S. Schlömer, G. Hansen, & M. Mastrandrea, Eds.). IPCC Working Group III Technical Support Unit, Potsdam Institute for Climate Impact Research. Geneva: IPCC

IPCC (Intergovernmental Panel on Climate Change). (2013a). Climate Change 2013: The physical science basis. In *Contribution of Working Group I to the Fifth Assessment Report of the Intergovernmental Panel on Climate Change*. Cambridge, UK and New York: Cambridge University Press.

IPCC (Intergovernmental Panel on Climate Change). (2013b). Summary for policymakers. In T. F. Stocker, D. Qin, G.-K. Plattner, M. Tignor, S. K. Allen, J. Boschung, A. Nauels, Y. Xia, V. Bex, & P. M. Midgley (Eds.), *Climate Change 2013: The physical science basis. Contribution of Working Group I to the Fifth Assessment Report of the Intergovernmental Panel on Climate Change*. Cambridge: Cambridge University Press.

IPCC (Intergovernmental Panel on Climate Change). (2013d). *Government and expert comments on draft Working Group 2 report*, Chapter 19. Spreadsheet 'WGIIAR5_SODCh19_annotation.pdf' previously available from IPCC website. Download available from the author.

IPCC (Intergovernmental Panel on Climate Change). (2013e). *Expert review comments on draft Working Group 3 first order draft report*. Spreadsheet available in pdf format from IPCC website as 'Expert Review Comments on the IPCC WGIII AR5 First Order Draft—Chapter X.pdf'. Retrieved January 20, 2019, from https://archive.ipcc.ch/pdf/assessment-report/ar5/wg3/drafts/ipcc_wg3_ar5_sod_comments_all-report.pdf

IPCC (Intergovernmental Panel on Climate Change). (2014a). Climate Change 2014: Synthesis report. In Core Writing Team, R. K. Pachauri, & L. A. Meyer

(Eds.), *Contribution of Working Groups I, II and III to the Fifth Assessment Report of the Intergovernmental Panel on Climate Change*. Geneva: IPCC.
IPCC (Intergovernmental Panel on Climate Change). (2014c). Climate Change 2014: Mitigation of climate change. In *Contribution of Working Group III to the Fifth Assessment Report of the Intergovernmental Panel on Climate Change*. Cambridge, UK and New York: Cambridge University Press.
Izrael, Y. A., Zakharov, V. M., Petrov, N. N., Ryaboshapko, A. G., Ivanov, V. N., Savchenko, A. V., et al. (2009). Field experiment on studying solar radiation passing through aerosol layers. *Russian Meteorology and Hydrology, 34*, 265–273.
Jackson, B. (2011, December 5). Gingrich on climate change. Retrieved January 9, 2019, from http://www.factcheck.org/2011/12/gingrich-on-climate-change/
Keith, D. W. (2013). *A case for climate engineering*. Cambridge, MA: MIT Press.
Keith, D. W. (2015). Climate engineering, no longer on the fringe. Interview 18 February. Available at Harvard School of Engineering and Applied Sciences website. Retrieved January 9, 2019, from http://www.seas.harvard.edu/news/2015/02/climate-engineering-no-longer-on-fringe
Keith, D. W., Parson, E., & Morgan, M. G. (2010). Research on global sun block needed now. *Nature, 463*, 426–427.
Kemp, M. (2005). Science in culture: Inventing an icon. *Nature, 437*(7063), 1238.
Kiehl, J. T. (2006). Geoengineering climate change: Treating the symptom over the cause? *Climatic Change, 77*(3), 227–228.
Kintisch, E. (2010). *Hack the planet: Science's best hope—Or worst nightmare—For averting climate catastrophe*. Hoboken, NJ: Wiley.
Kirby, A. (2003, December 4). Russia's climate tussle spins on. *BBC News*. Retrieved January 9, 2019, from http://news.bbc.co.uk/2/hi/science/nature/3288683.stm
Lane, L., & Bickel, J. E. (2013). *Solar radiation management: An evolving climate policy option*. Washington, DC: American Enterprise Institute.
Lane, L., Caldeira, K., Chatfield, R., Langhoff, S. (2007). *Workshop report on managing solar radiation*. NASA Ames Research Centre & Carnegie Institute of Washington, Moffett Field, CA, 18–19 November. Hanover, MD: NASA. (NASA/CP-2007-214558).
Lempert, R. J., & Prosnitz, D. (2011). *Governing geoengineering research: A political and technical vulnerability analysis of potential near-term options*. Santa Monica, CA: RAND Corporation.
Levitt, S. D., & Dubner, S. J. (2009). *Superfreakonomics*. New York: Harper Collins.

Liverman, D. M. (2009). Conventions of climate change: Constructions of danger and the dispossession of the atmosphere. *Journal of Historical Geography, 35*, 279–296.
Lovelock, J. (2008). A geophysiologist's thoughts on geoengineering. *Philosophical Transactions of the Royal Society A, 366*(1882), 3883–3890.
Lovelock, J. (2009). *The vanishing face of Gaia: A final warning.* London: Penguin Books.
MacMartin, D. G., Ricke, K. L., & Keith, D. W. (2018). Solar geoengineering as part of an overall strategy for meeting the 1.5°C Paris target. *Philosophical Transactions of the Royal Society A, 376*, 20160454.
Macnaghten, P., & Owen, R. (2011). Environmental science: Good governance for geoengineering. *Nature, 479*(7373), 291–292.
Macnaghten, P., & Szerszynski, B. (2013). Living the global social experiment: An analysis of public discourse on solar radiation management and its implications for governance. *Global Environmental Change, 23*, 465–474.
Markusson, N. (2013). Tensions in framings of geoengineering: Constitutive diversity and ambivalence. *Climate Geoengineering Governance Working Paper Series*: 003.
Markusson, N., Ginn, F., Ghaleigh, N. S., & Scott, V. (2014). 'In case of emergency press here': Framing geoengineering as a response to dangerous climate change. *WIREs Climate Change, 5*(2), 281–290.
Moore, J. C., Chen, Y., Cui, X. F., Yuan, W. P., Dong, W. J., Gao, Y., et al. (2016). Will China be the first to initiate climate engineering? *Earth's Future, 4*(12), 588–595.
Morton, O. (2007). Climate change: Is this what it takes to save the world? *Nature, 447*(7141), 132–136.
National Research Council (NRC). (2003). *Critical issues in weather modification research.* Washington, DC: Committee on the Status and Future Directions in U.S Weather Modification Research and Operations, National Academies Press.
National Research Council (NRC). (2015). *Climate intervention: Reflecting sunlight to cool Earth.* Washington, DC: National Academy of Sciences.
Nerlich, B., & Jaspal, R. (2012). Metaphors we die by? Geoengineering, metaphors, and the argument from catastrophe. *Metaphor and Symbol, 27*(2), 131–147.
Nordhaus, W. D. (1994). *Managing the global commons: The economics of climate change.* Cambridge, MA: MIT Press.
Nurse, P. (2011, September 8). I hope we never need geoengineering, but we must research it. *The Guardian.* Retrieved January 9, 2019, from https://

www.theguardian.com/environment/2011/sep/08/geoengineering-research-royal-society
Oldham, P., Szerszynski, B., Stilgoe, J., Brown, C., Eacott, B., & Yuille, A. (2014). Mapping the landscape of climate engineering. *Philosophical Transactions of the Royal Society A, 372*(2031), 1–20.
Oreskes, N. (2004). Beyond the Ivory Tower: The scientific consensus on climate change. *Science, 306*(5702), 1686.
Oschlies, A., & Klepper, G. (2017). Research for assessment, not deployment, of climate engineering: The German Research Foundation's Priority Program SPP 1689. *Earth's Future, 5*(1), 128–134.
Pierrehumbert, R. (2015, February 10). Climate hacking is barking mad. *Slate*. Retrieved January 9, 2019, from http://www.slate.com/articles/health_and_science/science/2015/02/nrc_geoengineering_report_climate_hacking_is_dangerous_and_barking_mad.html
Porter, K. E., & Hulme, M. (2013). The emergence of the geoengineering debate in the UK print media: A frame analysis. *The Geographical Journal, 179*(4), 342–355.
Ricke, K., Morgan, M. G., Apt, J., Victor, D., & Steinbruner, J. (2008). *Unilateral geoengineering: Non-technical briefing notes for a workshop at the Council on Foreign Relations*, Washington, DC, May 5, 2008. Retrieved January 9, 2019, from http://www.cfr.org/content/thinktank/GeoEng_Jan2709.pdf
Rickels, W., Klepper, G., Dovern, J., Betz, G., Brachatzek, N., Cacean, S., et al. (2011). *Large-scale intentional interventions into the climate system? Assessing the climate engineering debate*. Scoping report conducted on behalf of the German Federal Ministry of Education and Research (BMBF), Kiel Earth Institute.
Rittel, H., & Webber, M. (1973). Dilemmas in a general theory of planning. *Policy Sciences, 4*, 155–169.
Robine, J.-M., Cheung, S. L., Le Roy, S., Van Oyen, H., Griffiths, C., Michel, J. P., et al. (2008). Death toll exceeded 70,000 in Europe during the summer of 2003. *Comptes Rendus Biologies, 331*(2), 171–178.
Robock, A. (2008). 20 reasons why geoengineering may be a bad idea. *Bulletin of the Atomic Scientists, 64*(2), 14–18.
Robock, A. (2014, May 5). A case against climate engineering. *Huffington Post*. Retrieved January 9, 2019, from https://www.huffingtonpost.com/alan-robock/a-case-against-climate-engineering_b_5264200.html

Robock, A., Marquardt, A. B., Kravitz, B., & Stenchikov, G. (2009). Benefits, risks, and costs of stratospheric geoengineering. *Geophysical Research Letters, 36*(19), 1–9.

Royal Society. (2009). *Geoengineering the climate: Science, governance and uncertainty*. RS Policy document 10/09. London: Royal Society. Retrieved January 9, 2019, from https://royalsociety.org/~/media/Royal_Society_Content/policy/publications/2009/8693.pdf

Schelling, T. C. (1996). The economic diplomacy of geoengineering. *Climatic Change, 33*(3), 303–307.

Schiermeier, Q. (2003). Climate change: The long road from Kyoto. *Nature, 426*(6968), 756.

Schmidt, G. (2006, June 28). Geo-engineering in vogue. *RealClimate*. Retrieved January 9, 2019, from http://www.realclimate.org/index.php/archives/2006/06/geo-engineering-in-vogue/

Schnare, D. W. (2007). *A framework to prevent the catastrophic effects of global warming using solar radiation management (geo-engineering)*. Supplement to Testimony before the United States Senate Committee on Environment and Public Works.

Schneider, S. (2001). Earth systems engineering and management. *Nature, 409*(6818), 417–421.

Schrickel, I. (2014). Images of feasibility: On the viscourse of climate engineering. In B. Schneider & T. Nocke (Eds.), *Image politics of climate change: Visualizations, imaginations, documentations*. Bielefeld: Transcript Verlag.

Schwartz, P., & Randall, D. (2003). An abrupt climate change scenario and its implications for United States national security. Washington, DC: U.S. Department of Defense. Retrieved from http://catalogue.nla.gov.au/Record/3834225 or http://eesc.columbia.edu/courses/v1003/readings/Pentagon.pdf

Shaw, C., & Nerlich, B. (2015). Metaphor as a mechanism of global climate change governance: A study of international policies, 1992–2012. *Ecological Economics, 109*, 34–40.

Sikka, T. (2012). A critical discourse analysis of geoengineering advocacy. *Critical Discourse Studies, 9*, 163–175.

Singer, P. (2002). *One World: The ethics of globalization*. New Haven, CT: Yale University Press.

Sinitsyna, T. (2008, June 25). Ways to tame the climate. Opinion piece in *Sputnik News*. Retrieved January 9, 2019, from http://sputniknews.com/analysis/20080625/112056346.html

Stilgoe, J. (2015). *Experiment Earth: Responsible innovation in geoengineering.* Abingdon: Routledge.
Teller, E. (1997, October 17). Sunscreen for planet earth. *Wall Street Journal.* Retrieved January 9, 2019, from http://www.hoover.org/research/sunscreen-planet-earth-0
Teller, E., Hyde, R., & Wood, L. (2002). *Active climate stabilization: Practical physics-based approaches to prevention of climate change.* National Academy of Engineering Symposium, Washington, DC, April 23–24. Preprint UCRL-JC-148012.
Tyndall Centre Cambridge-MIT Institute Symposium. (2004). Macro-engineering options for climate change management and mitigation.
UNFCCC (United Nations Framework Convention on Climate Change). (2015). *Adoption of the Paris agreement.* FCCC/CP/2015/L.9/Rev.1. Conference of the Parties, Paris, December 12.
US House of Representatives Committee on Science and Technology (USHCST). (2009). *Hearing: Geoengineering: Parts I, II and III, assessing the implications of large scale climate intervention* (pp. 111–162, 111–175, and 111–188). Washington, DC: US House of Representatives. Retrieved January 20, 2019, from https://www.govinfo.gov/content/pkg/CHRG-111hhrg53007/pdf/CHRG-111hhrg53007.pdf
US House of Representatives Committee on Science and Technology (USHCST). (2010). *Engineering the climate: Research needs and strategies for international coordination.* Report by Bart Gordon. Retrieved January 20, 2019, from https://science.house.gov/sites/democrats.science.house.gov/files/10-29%20Chairman%20Gordon%20Climate%20Engineering%20report%20-%20FINAL.pdf
Van Hemert, M. (2017). Speculative promise as a driver in climate engineering research: The case of Paul Crutzen's back-of-the-envelope calculation on solar dimming with sulfate aerosols. *Futures, 92,* 80–89.
Victor, D. G. (2011). *Global warming gridlock: Creating more effective strategies for protecting the planet.* Cambridge: Cambridge University Press.
Wagner, G., & Weitzman, M. L. (2015). *Climate shock.* Princeton, NJ: Princeton University Press.
White House. (2001). President Bush discusses global climate change. Press Release, June 11. Retrieved January 9, 2019, from http://georgewbush-whitehouse.archives.gov/news/releases/2001/06/20010611-2.html
Wigley, T. M. L. (2006). A combined mitigation/geoengineering approach to climate stabilization. *Science, 314,* 452–454.

4

Competing Imaginaries of Solar Geoengineering

Since solar geoengineering (SGE) does not yet exist as a proven or deployed technology, we need to engage with the *idea* of SGE, with *how it is imagined*, and the *imagined world* in which it is expected to be deployed. Indeed, how SGE is imagined will likely shape the form the technology takes and whether it emerges as a deployed technology at all.

In this Chapter I identify and explore three competing sociotechnical imaginaries of SGE and the worlds they envisage. I label these the Un-Natural, the Chemtrail, and the Imperial imaginaries. The first regards SGE as an illegitimate intervention in the natural world, a plutocratic initiative deploying a dangerous technology of hubris (a techno-fix), a disaster for the world's poorest, and one which should not be allowed to proceed. The second, Chemtrails, sees SGE as part of an elite conspiracy to actively, and secretly, poison the land, air and water by spraying toxic chemicals—'chemtrails'—from airplanes for a variety of malevolent purposes. The third, the Imperial imaginary, imagines governing the climate as a component of governing the world and tackling the risks of climate change. It binds together, not always comfortably, three subsidiary narratives: a Market narrative, a Geo-management narrative and a Salvation narrative.

The first and second of these "sociotechnical imaginaries," to use Sheila Jasanoff's term (2015), are clearly oppositional with regard to SGE, whilst the last is broadly supportive of a turn to SGE. The Chemtrails imaginary operates in a different register to the others. It has been called a conspiracy theory and its truth claims are regarded by scientists as unfounded (Shearer et al. 2016). I include it since my focus is on what worlds are imagined, rather than the truth or falseness of the claims being made, and this requires interrogating all knowledge claims symmetrically (Bloor 1991). As Pelkmans and Machold put it: "assessments of conspiracy theories should focus not on the epistemological qualities of these theories but on their interactions with the socio-political fields through which they travel" (2011: 66).

Here I examine the thinking and argumentation which animates these three competing accounts of what SGE is, what it is for, where it fits in to broader understandings of the world, and what its future trajectory ought to be. I will explore the assumptions, stated and implicit, of each imaginary: assumptions about the contemporary world, existing structures of political and economic power, nature and the environment, technology and science, democracy and climate change, as well as their stance towards the future, and their 'mood.' Which, if any, of these competing imaginaries prevails will shape, at least as much as the realities of rising temperatures, whether SGE moves from an imagined to a deployed technology. Whilst the nascent Imperial imaginary is preponderant in the corridors of power it is struggling to become hegemonic. It is difficult, precisely because none is hegemonic, to extract these imaginaries directly from the major institutional assessment reports, although elements of the Un-Natural and Imperial narratives are contained or implicit within each report. I therefore outline each imaginary, or associated narrative, by focusing on a representative text in an attempt to present each in their most articulate and coherent version.

Un-Natural: The Perils and Injustice of Geopiracy

They call it geoengineering—we call it geopiracy.—ETC Group (2010a)

4 Competing Imaginaries of Solar Geoengineering 125

The Action Group on Erosion, Technology and Concentration (known as the ETC Group) is a small Canadian-based global NGO which, according to its website, "works to address the socioeconomic and ecological issues surrounding new technologies that could have an impact on the world's poorest and most vulnerable people" (ETC Group n.d.) It campaigns against geoengineering testing, experimentation and deployment, and is part of a larger network called Hands Off Mother Earth (H.O.M.E.). ETC Group and H.O.M.E. have not directly shaped the institutional reports on SGE although they could be said to have influenced them indirectly: many of the reports reference ETC's views and some have included panelists holding views broadly similar, but expressed in less activist terms.

"Geopiracy" is how the ETC Group (2010a) characterizes geoengineering: a "gamble with Gaia" (2010b). The cover design for its 2010 report *Geopiracy: The Case Against Geoengineering* consists of an adaptation of Edvard Munch's well-known painting *The Scream*, said to reflect the painter's "feeling of 'a great unending scream piercing through nature'" (ETC Group 2010a: i). At stake in geoengineering is "international control of planetary systems: our water, lands and air," as well as retaining existing commitments to mitigation and adaptation. If this "quick, cheap fix" is adopted then rich governments will devote resources to "this 'scientific solution' and there will be no resources to help the global South fend off the chaos ahead" (ETC Group 2010a: 1). Further, there is no reason why the global South should "trust that the governments, industries or scientists of the biggest carbon-emitting states will protect their interests" (ETC Group 2010a: 4).

For the ETC Group, "a moratorium on real-world geoengineering experimentation is urgent", and the United Nations and the International Court of Justice should confirm that any experimentation would be in breach of the 1977 United Nations Convention prohibiting the hostile use of Environmental Modification Convention (ENMOD) (2010a: 2).

The ETC Group is keen to unveil the network of interests, plutocrats and scientists engaged in geoengineering. They talk of a "geo-clique," and list the scientists involved in research who have also lodged geoengineering-related patents, and the ultra-wealthy individuals who have engaged with geoengineering (such as Bill Gates). They state that "there is a complex web of connections between big capital and the global technofixers, comprised of researchers, multinational corporations and small start-ups, the military establishment and respected think tanks, policy makers and politicians" (ETC Group 2010a: 38). The ETC Group also draws attention to the extensive links of SGE's boosters to the US military and corporate establishment (ETC/Biofuelwatch 2017: 42–3). In the process they perhaps assume a greater commitment to geoengineering, at least to SGE, and a greater unity of purpose amongst such participants than may in fact be the case. This helps explain an occasionally conspiratorial tone which permeates their understanding of geoengineering's emergence. They make a more general point about elites in *The Big Bad Fix*, an update of their earlier report produced this time with another NGO, Biofuelwatch: "current debates about this big techno-fix are limited to a small group of self-proclaimed experts reproducing undemocratic worldviews and technocratic, reductionist perspectives", and exclude "developing countries, indigenous peoples, and local communities" (ETC/Biofuelwatch 2017: 4).

In exploring SGE in particular, which ETC Group calls "artificial volcanoes," whilst there are "many unknown impacts … already there is research" suggesting the likelihood of a range of negative effects. Here they draw on atmospheric climate scientist Alan Robock's paper '20 Reasons Why Geoengineering may be a Bad Idea' (2008) and mention effects including differential regional impacts, drought, ozone damage, the "whitening" of skies, and the expectation of a "bounce-back" effect after termination. ETC Group also note, as Robock does not, that "geoengineering the stratosphere makes it easier for industry to continue its own atmospheric pollution" (2010a: 26).

This concern with capitalism (although they do not name it) and with power relations (especially North-South) permeates the ETC Group's vision. "Should a 'Plan B' ever be agreed upon," patent applications being in private hands make the prospect of the technology being implemented "terrifying": "planet-altering technologies should never be undertaken for private profit" (ETC Group 2010a: 33). They are concerned too about "the self-serving gambit of climate-deniers and cringing politicians in the temperate zone" recasting geoengineering as "foreign aid" (ETC Group 2013b). In this they are critiquing, as we shall soon see, an element of the Imperial imaginary of SGE. As the ETC Group put it in their *Geopiracy* report: "Peoples of the South should be in control of climate response decisions instead of being cast as helpless victims waiting to be saved by the technologies of the North, with lip service to their interests the only acknowledgement of their dilemma" (ETC Group 2010a: 37). In their later report they express the fear that "the world is sleepwalking toward a geoengineered future" (ETC/Biofuelwatch 2017: 5), that this will "render humanity indefinitely dependent on a technocratic elite" (ETC/Biofuelwatch 2017: 6), and that the "would-be Emperors [the current leaders of China, Russia and the United States] could even form a 'climate coalition of the willing' using geoengineering to try to protect their part of the Northern hemisphere, with little regard for the rest of the world" (ETC/Biofuelwatch 2017: 8).

For the ETC Group, geoengineering is a manifestation of "scientific hubris," a "techno-fix" and a risky undertaking which "flies in the face of precaution" (2010a: 3). They lament an inability to learn the lessons from past weather modification activities.[1] In a neat turn of phrase they ask whether "the 'Hot Worriers' of today have learned from the 'Cold Warriors' of yesterday" (ETC Group 2013a: 5)? Their problem is with "Big Science" rather than science as such:

> We need a thousand candles of brilliant research rather than a new Manhattan Project. By definition, the practical responses to climate change must change with the latitudes and the altitudes and the ecosystems. ... "Big" Science is going to have to learn to become "diverse" science and to work with Southern governments, local communities, indigenous peoples and peasant farmers already trying to respond to this crisis. Humility will need to replace hubris. (ETC Group 2009: 35)

Pervading the ETC Group's account is a view that geoengineering is in some sense 'un-natural,' an unnecessary intervention into Gaia (ETC Group 2010b), the living, dynamic and inter-related Earth systems. This approach is manifest in the H.O.M.E. campaign sticker which depicts a splayed hand signaling 'stop,' positioned over the iconic 'blue marble'

---

[1] Mike Hulme's book *Can science fix climate change?* (2014) makes similar arguments in some respects when he concludes that SGE is "undesirable, ungovernable and unreliable". So too does James Fleming in *Fixing the Sky* (2010).

Fig. 4.1 Hands Off Mother Earth (HOME) sticker (Artwork: Shtig, courtesy of HOME Campaign)

image of Earth (see Fig. 4.1). The Hands Off Mother Earth *Manifesto Against Geoengineering* of October 2018 argues that geoengineering technologies and practices "violate the natural laws, creative principles and the Territorial Integrity of Mother earth and Father Sky". They "may disrupt local and regional weather patterns … with potentially catastrophic effects". They threaten global peace and security, depend on military infrastructure and have "significant potential to be weaponized". SGE is "dangerous, unnecessary and unjust" and "our home, lands and territories are not a laboratory for planetary-scale environmental modification technologies". (HOME 2018: n.p.)

The Un-Natural imaginary paints a gloomy and dystopian picture of what a geoengineered world would be like. It imagines SGE to be un-necessary, un-natural, hubristic, unreliable and risky, and driven by self-interest and a desire to avoid mitigation. It is against the interests of the global South. In its aim to buy time it places "an unjust burden on future generations" (ETC/Biofuelwatch 2017: 31). It is a technology that should be highly regulated, even stopped. In aiming simply at "weakening some of the symptoms of climate change" (HOME 2018: n.p.) it detracts from tackling the climate crisis.

Imperial: Acclimatising the World

I label the second widespread imaginary of SGE the Imperial imaginary. It is an imaginary which is favourable to the further development and deployment of SGE, notwithstanding that this advocacy is currently expressed cautiously and sometimes has reservations attached. It is most eloquently expressed by individuals close to the centres of power—economic, political, scientific, academic and military—and the existing establishment, especially in the United States. It currently exists in three distinct flavours, narratives which are slightly different, albeit complementary to each other: a Market narrative enthusiastic about SGE's potential to cool the planet cheaply; a Geo-management narrative which sees SGE as a powerful instrument, to be wielded by geo-politically powerful nations, in the face of a terrible climate prognosis and a politically unstable world; and a Salvation narrative seeing the use of some SGE as essential to save humanity from the ravages of climate change. The views of their proponents have been especially influential in the major institutional reports and in shaping elite discourses around geoengineering. It is therefore important to unpack the differing logics of each of the component narratives of the Imperial imaginary.

Despite their differences, what unites these narratives is both a favourable disposition towards SGE and a number of assumptions they hold in common. These include assumptions about the centrality of technological solutions to climate change, about the desirability (or inevitability) of the current order of market globalism, about the ability to physically manage the climate system, and about the entitlement of the powerful to do so. I will look at each of the component narratives of the Imperial imaginary in turn.

Market: A Techno-Fix to Enable Business-as-Usual

> If we could come up with a geoengineering answer to this problem, then Copenhagen wouldn't be necessary. We could carry on flying our planes and driving our cars.—Sir Richard Branson (cited in Revkin 2009)

It is in the writings of Eric Bickel and Lee Lane that the Market narrative is most eloquently and comprehensively expressed. Bickel and Lane have been the lead authors of a number of think tank reports on geoengineering, most notably those of the American Enterprise Institute (AEI) (Lane and Bickel 2013), and the Copenhagen Climate Consensus (CCC) (Bickel and Lane 2012). Lane was also a co-author of the NASA report on geoengineering (Lane et al. 2007).

The main elements of their narrative can be summarised as follows. On climate change and existing climate policy they argue that greenhouse gas controls have failed and "the prospect of an effective GHG control policy remains far in the future, and actual impacts on climate much more distant still" (Lane and Bickel 2013: 1). In effect, in this argument, there is no reliance on an emergency discourse to justify SGE, and mitigation has not only failed, but was *never* desirable as the primary policy goal. Some climate risk argument is implicit in the AEI report, but it is not especially central since the climate effects are seen as more temporally distant. However, in a report for the CCC, SGE's ability to cool rapidly, Bickel and Lane argue, allows it "to play an important risk management role" (Bickel and Lane 2012: 4). The main expectation is that people will adapt their "customs, dress, crops, structures, locations, and practices to a changing climate and its effects" (Bickel and Lane 2012: 3). However, adaptation is expensive and will become more so "if the change is too rapid or too large" (Bickel and Lane 2012: 1). The attraction of SGE, therefore, is that it may slow the pace of climate change and lessen "both the harm from climate change and the costs of adapting to it" (Bickel and Lane 2012: 1). "It represents a possible 'force multiplier' for adaptation efforts" (Bickel and Lane 2012: 5). The argument that SGE is cheaper than mitigation by an order of magnitude is strongly embraced and makes SGE especially attractive to the Market narrative.

Technology is the essential component in the Bickel and Lane argument. The ideal is if "green energy technologies" become so cheap that emission controls become "politically more palatable or even superfluous" (Bickel and Lane 2012: 2). But SGE is a stopgap in the absence of this and it is a technology "believed to be well within our current capabilities" (Bickel and Lane 2012: 20). Bickel and Lane regret that much of the criticism of SGE is "tinged with moral censure," including the charge that government has not achieved enough mitigation, as well as suggestions that the enterprise itself is an example of hubris (Bickel and Lane 2012: 15).

Nature does not feature in Bickel and Lane's argument. However they do accuse environmental NGOs with seeing "any human interference with nature as morally wrong" and charge them with being "largely deaf to the concept of instrumental rationality" (Bickel and Lane 2012: 17). Democracy and any consideration of how SGE and SGE research might be governed is also absent from their account, apart from one intriguing reference: "It is an interesting question whether democratic forms of government can conduct an R&D program on a concept as polarizing as SRM is at the present time" (Lane and Bickel 2013: 12). This suggests they are alert to, and largely untroubled by, the argument that SGE and democracy may not be compatible. Critics of SGE are charged with having a dystopic view of the technology and its ills (Bickel and Lane 2012: 15). But the imaginary implicit in Bickel and Lane's argument is not particularly positive or visionary, perhaps because it simultaneously tries to present climate change as not especially urgent or dramatic, whilst being enthusiastic about the potential of SGE and persistently downplaying many widely-acknowledged risks.

The tone of the Market narrative is matter of fact, and its arguments constitute a type of market fundamentalism, where entrepreneurialism, growth and efficiency are the highest societal purpose. Indeed, the main focus of the work is to model the cost-effectiveness of SGE as compared to mitigation.[2] It is simultaneously an argument for SGE and an argument to de-emphasize mitigation as a policy preference. This account helps explain why some of the enthusiasts for SGE come from the ranks of people and institutions on the US political right, often labeled 'climate denialist.' So, for example, we can find David Schnare of the Heartland Institute (notorious for their annual conferences on de-bunking climate change) arguing for SGE on the grounds it will be quicker and cheaper (Schnare n.d.).[3] More recently, in November 2017, the Republican chairman of the

---

[2] Bickel has also co-authored with Agrawal a 2013 paper vigorously challenging the economic modelling of Goes et al. (2011). This is significant because Goes, Tuana and Keller conclude, using the same style of economic modelling but different assumptions, that SGE would only be economically rational in very limited (and highly unlikely) circumstances.

[3] Schnare is a prominent advisor to Donald Trump on climate policy and restructuring the US Environmental Protection Agency.

Energy Subcommittee of the House Committee on Science, Space and Technology in the US Congress, Randy Weber, commented that "the future is bright for geoengineering". Weber has previously grilled scientists on "global cooling" and rejected a carbon pollution tax as "blasphemy" (Ellison 2018: n.p.).

In the right-wing narrative solar geoengineering is attractive for those alert to some of the implications of acknowledging climate change as a serious issue: namely that contemporary patterns of consumption, the ideology of limitless growth, and the 'take-make-waste' structure of production may be implicated. The Market narrative acknowledges that climate change is happening, although it may question this being a 'crisis'. Its proponents also have a primary commitment to avoiding policies which are expensive or market-constraining, or which may limit economic growth and 'development' in any way. On this interpretation Market proponents of SGE are best understood as 'climate inactivist' rather than 'climate denialist'. It is this line of thinking which leads to SGE being embraced by those hostile to existing mitigation policies. It allows them to be scientifically literate on the issue of climate without conceding their core ideology. It is attracted by observations, such as that of eminent economist William Nordhaus that "geoengineering is at present the only economically competitive technology to offset global warming" (2007: n.p.).

The embrace of business-as-usual capitalism is central to the Market narrative and such views are not confined to the fringes. They can be found in comments by the Nobel laureate economist, Thomas Schelling, speaking at an AEI event: that one advantage of geoengineering "is that it does not involve drastically changing the way billions of people live and care for themselves on a daily basis." (2009). Leading US politician and Donald Trump ally Newt Gingrich is also an enthusiast for SGE (2008). Similar interest in geoengineering has been expressed by another Trump advisor on climate and environmental policy, Myron Ebell (2009). The Market narrative is not limited to the US political right. It is essentially the perspective expressed by the more liberally-inclined Richard Branson at the launch of his Virgin Climate Challenge in 2009 cited above. And, as I have shown, even the other strands of the Imperial imaginary commonly emphasise SGE's purported cheapness although they do not give this argument as much prominence as the Market narrative does. SGE's virtue, in the Market narrative, is that it addresses climate change cheaply whilst enabling the fundamentals of the dominant global economic order to continue.

Geo-management: Taking Charge of the Climate Crisis

> Even if a large dose of geoengineering is applied, quite a lot of triage may remain as well.—David Victor (2011: 198)

David Victor is an influential and respected academic figure in the shaping of climate and geoengineering policy, especially in the US. I take his views as an exemplar of the Geo-management narrative. Victor has been involved as panelist or reviewer in a number of the most influential institutional reports on geoengineering (for example NRC 2015a, b, c, d; BPC 2011; IPCC 2012; GAO 2010). His support for SGE, as found in his monograph *Global Warming Gridlock*, is rooted in a critical account of the then-existing Kyoto Treaty climate policy with its focus on mitigation and adaptation, targeted reductions by 'developed' countries, and climate multilateralism. He examines these issues through the lens of international relations and geo-politics. He is especially influential in articulating what he calls the 'hard-nosed politics' of climate change including the need to develop and probably deploy SGE.

For Victor, the existing UN approach to climate policy has not worked because "it involves too many countries and issues" (2011: xxviii). Better to focus on a "club" approach starting with the few "countries that matter," before expanding the club through accession agreements, in the manner of the WTO. For Victor, any ambition to stay below 2°C through mitigation is "probably history" and "we" need to be "bracing for change" (2011: xxxii). Adaptation is locally important, but generally not an internationally-effective instrument, and it should largely be left up to each country. Victor acknowledges that adaptation is insufficient or non-existent in many countries, notably those that are not rich enough or "smart enough" to adapt, or have corrupt governments. The expressions "rogue states," "failed states," "weak states" and "small states" occur frequently in the book and appear to be what Victor has in mind. This is the language of policy think-tanks in the global North. It is hard not to see in this an imperial mindset: these are countries for whom (and to whom?) things must be done. New technology is therefore critical: "technology, not castor oil [i.e. bitter medicine], is how most environmental problems get solved" (Victor 2011: xvii). Further, only about ten countries "matter" when it comes to innovation (Victor 2011: xxxii).

For Victor this means geoengineering (by which he means mainly SGE) and "triage" (a term derived from emergency medicine), are central "bracing for change" strategies. Triage, for Victor, is for situations where adaptation has proven impossible, or too expensive, or where there is not much worth protecting in monetary terms (2011: 185). It may entail "walking away" (Victor 2011: 185). What is meant by this innocuous phrase is not elaborated, but the term "triage" is suggestive, including as it does the notion of abandoning those who are in any event likely to die. Victor says new policy options should also be looked into, including "for countries that have consistently failed to create the right context for economic development (and adaptation)." He mentions two such options: migration and "receivership" (Victor 2011: 185). By the latter he presumably means re-colonization or resumption of trusteeship of some sort.

Alongside Triage, and possibly to help reduce the need for it, sits SGE. For Victor, in preparing for an extreme climate future, and possible climate emergencies, where adaptation is impractical and costly, we need to develop the ability to "mask the horrors with geoengineering" (Victor 2011: 185). 'We' need to research and test SGE and governments should fund this. We need to make the public more comfortable with this option. We need to be able to use it as part of a "cocktail" of interventions in much the way we combat AIDS. For example, since SGE does not address ocean acidification, that problem may need a different geoengineering intervention as part of the 'cocktail' of climate interventions.

For Victor, working on SGE only with prior international agreement, as some suggest, is misguided and would result in "gridlock." Any conceivable treaty would only amount to a ban. Making SGE "taboo" would be the worst policy because it is "likely to be most constraining on the countries (and their subjects) who are likely to do the most responsible testing, assessment, and (if needed) deployment of geoengineering systems" (Victor 2011: 193). Rather, "A better understanding that a taboo is a dangerous policy because it cedes defeat in this race will help realign the politics in the liberal democratic countries in favor of supporting a research program" (Victor 2011: 196).

Victor's use of terms such as "subjects," "triage," "receivership," and "countries that matter," is revealing. This is the world imagined from the standpoint of the geo-politically powerful. There is a presumption that already powerful countries and interests in the world should manage climate and take decisions for the "rest," including developing and deploying SGE. But it is hard to imagine that SGE imposed by a club of the powerful would be deployed benevolently, in ways good for most but possibly inconvenient to the climate and temperature desires of the powerful. As Stilgoe has pointed out, history suggests that "centralised sociotechnical systems controlled by rich people, will tend to exacerbate the gap between rich and poor rather than close it" (2015: 190). And it is hard to see why the majority of the world's population, in the global South and in countries that do not "matter," would accede to this vision of SGE.[4]

Victor seems to recognise this when he urges that any attempts to regulate SGE multilaterally should be blocked. But this then leaves him with the problem of how to make SGE acceptable. The approach he suggests is to normalize SGE "through experience and dialog". "Meaningful norms are not crafted from thin air," Victor argues. They must "make sense to pivotal players," "provide a useful function" and then "become socialized through practice." He believes that the spread of human rights norms may be a model (Victor 2011: 195). Certainly, although he does not acknowledge it, this would require a framing shift away from his own rather despairing account of climate change. It would suggest arguing that 'we are doing SGE for you, not for us.'

Nature and the environment are only present indirectly in Victor's imaginary. He laments that environmental scholars "care too much about their subject," and are too "green" and "evangelical" around "the need to halt planetary destruction" (Victor 2011: x–xi). In his account, nature's value appears to be mainly utilitarian and its going out of kilter, such as

---

[4] This tension can only be exacerbated in a context where the United States withdraws from multilateral climate agreements whilst remaining the country most wedded to developing SGE programs.

if climate 'tips,' is of concern because of the risks and impacts on human society. Nature is to be managed. We should adapt to environmental problems where possible, or find ways of masking them (such as by SGE) where needed or if the problems become extreme (Victor 2011: 199). Victor acknowledges that "masking requires the cooperation of nature, which is harder to predict and usually less pliable" (2011: 199).

Victor imagines a miserable present and an even more miserable and dystopian future, one which 'we' need to get used to. He concludes his book: "Scenarios where geoengineering systems are deployed are not far off; getting ready for them probably means starting to test and evaluate systems today. The Brave New World is here" (Victor 2011: 200). The allusion, of course, is to Aldous Huxley's famous techno-dystopian novel, where ten World Controllers run a benevolent dictatorship, a World State. In the novel, this dictatorship controls most of the world apart from a few inconvenient areas where the 'savages' are left to their own devices.

In short, for Victor, the economic and geo-political *status quo* and their maintenance is largely assumed (Victor 2011: xvii). SGE is imagined as an elite project, an additional leg of climate policy, and part of suite of adjustments and additions to current approaches to climate change. Climate change is a serious problem for which big technologies are a central solution, and SGE may require undemocratic 'club' approaches to be brought into being.

Salvation: Saving the World from Climate Risk

> It is hyperbolic but not inaccurate to call [solar geoengineering] a cheap tool that could green the world.—David Keith (2013: x)

The Imperial imaginary's Salvation narrative presents SGE as essential to save the world from the risks of climate change. I take David Keith's short book, *A case for climate engineering*, as the strongest version of this argument. He acknowledges, in the book's title, that this is *a* case for SGE, rather than claiming it to be *the* case. Keith is especially influential in the construction of SGE both as an idea and in practice. He has published

4 Competing Imaginaries of Solar Geoengineering 137

extensively on the topic over almost three decades, and is actively involved in plans to conduct outdoor experiments through the SCoPeX project. He is a leading promoter of SGE research, has a significant media presence commenting on SGE, and he has directly shaped and influenced *all* the major institutional reports.[5] His book allows him to express his case for SGE without the constraints of reaching consensus which inevitably shape any institutional policy report—whose panels have generally also included members sceptical of the geoengineering turn. Here I will focus on summarizing the explicit and implicit imaginary contained in his book, and occasionally reference related literature by Keith.

Keith regards predictions of climate catastrophe as "rhetorical flourishes" and holds that "there is no imminent existential threat" (2013: 24). However, "climate risks are serious" and the rate of change is especially concerning since our infrastructure, crops and coastal cities have evolved for the current climate (Keith 2013: 24–7). The chances of stopping before 500 ppm or holding global temperature rise below 2 degrees are seen as remote. Keith's imagining of how climate change will unfold is consistent with successive IPCC science assessments. This aspect of the Salvation narrative is distinct from the Market narrative outlined above which downplays the seriousness of climate change.

Keith, like Victor, imagines geoengineering as one of a number of strategies to address climate change, and sees it as a third leg of climate management, alongside mitigation and adaptation (Olson 2011: 4). Keith does not regard SGE as benign or even desirable: "It's a terrible option" (quoted in Wagner and Weitzman 2012: n.p.). He even sees the "intuitive revulsion" towards SGE as "healthy" since "our gadget-obsessed culture is all too easily drawn to a shiny new tech fix" (Keith 2013: xi). However, Keith does see SGE as necessary to reduce climate risk and save the poor. Keith is especially exercised by "the ugly prospect of rich people arguing that we should reject the geoengineering Band-Aid—thus denying what

---

[5] Keith was a member of the Royal Society (2009) working group, on the scientific committee of the IPCC Expert Group and the invited keynote speaker for its 2011 session in Peru (IPCC 2012: 19), as well as a reviewer of the NRC report (NRC 2015a, b, c, d). He was also a co-author of the NOVIM report (Blackstock et al. 2009), "helped fine tune several of the [Woodrow Wilson] report's recommendations" (Olson 2011: vii), was on the Bipartisan Policy Centre's task force (BPC 2011: 4), on the scientific panel of the GAO report (2010), on the working group of SRMGI (2011), and was one of four witnesses to testify to Congress about geoengineering (USHCST 2010: 4).

may be a large benefit to the poor—in order to goad the rich into cutting emissions" (2013: 137).[6] For him, this "repugnant" argument comes from rich and educated first-worlders claiming to speak for the poor. Better, he suggests, to geoengineer as a pro-poor poverty reduction strategy, not to mention a "greening" one. Climate change, and resultant crop losses, "will put millions of the poorest at risk" (Keith 2013: 138). Whilst the emissions and energy use come mainly from the rich, "the burdens of climate change fall most strongly on the poor who will also benefit most strongly from geoengineering (if it works) when it reduces these burdens" (Keith 2013: 139).

Keith cautions against rushing to use SGE and advocates using it to offset only some, not all, warming relative to pre-industrial temperatures (2013: xii). He imagines "the wise use of geoengineering," one which recognizes the potential risks of the technology (Keith 2013: 174). The obvious question which follows is 'who will ensure wise use?' Keith attaches great importance to finding a suitable, and plausibly democratic, model of governance, although he does not develop one.[7]

Keith regards the commonly-cited risks of SGE as overstated. He challenges, for example, the widespread claim that SGE would disrupt the South Asian monsoon, obviously a major negative for any SGE proponent. His choice of words is precise, and includes many qualifiers: "…*all* studies to date … suggest that the *appropriate* use of geoengineering *could* reduce climate risks to *Asian* agriculture" (Keith 2013: 58 first emphasis in the original, remainder my own). But he mainly contrasts any risks with the far greater risks (as he sees it) of unmitigated climate change (Keith 2013: 72). In this argument, any possible dangers of hubris are outweighed by the dangers of climate change and its attendant risks.

For Keith, SGE is an experimental technology, and it is more engineering than science (Keith 2013: 116). He emphasizes that SGE is a mundane and

---

[6] The implication here is that the risk of diluting mitigation efforts—so-called "moral hazard"—is the major objection marshalled by SGE's opponents.

[7] In this regard the Salvation and Geo-management narratives can be seen as complementary. For Victor, *realpolitik* suggests a small club of states agree on SGE governance, in the apparent belief that wise (or wise-enough) use will follow if a 'responsible' state leads. It should be noted that a number of major initiatives around governing SGE have failed to produce agreement that testing and trial deployment should proceed, and that a major new attempt to address governance is currently underway: the Carnegie Climate Geoengineering Governance Initiative (C2G2 2018).

not especially innovative technology.⁸ The aim of deploying SGE is to 'shave the peaks' off global warming or slow the rate of warming, and allow time for mitigation to occur and for carbon dioxide removal technologies to be developed and deployed. He implicitly rejects the argument made by others that embarking on SGE is a centuries-long or millennial commitment. He imagines starting with small injections, tracking the effects, and then slowly ramping the dosages up or down so there is no single shock to the system and reversibility is possible. The risks of doing this, he argues, are minimal. Keith's goal is akin to the "climate stabilization" argument proposed by Wigley (2006). But here the aim is not to restore the Earth to a previous temperature but to manage a transition to a new climate and, ideally, get to a point where the SGE intervention can cease.

When it comes to nature, Keith argues that his interest in SGE is "rooted in a concern that environmentalism has lost its way" (2013: xv). He proclaims a love of the outdoors and a suspicion of cost-benefit utilitarian claims about the value of nature. "The challenge is to craft an environmental ethic that recognizes non-utilitarian values in the natural world without asserting that these values trump all others and without making naïve claims of a sharp distinction between nature and civilization" (p. xvii). In doing so he engages with contemporary thinking about nature, influenced by Marris's work *Rambunctious Garden* which argues, says Keith, that "environmentalists should abandon the obsessive defense of pristine nature in favor of an expanded environmental ethic that embraces the messy but vibrant reality of landscapes shaped by human action" (p. xviii). Recognizing humanity's role in shaping landscapes does not "drain them of value or turn them into an artifact." Indeed, their value lies in their history, "the co-evolution of nature, culture and technology" (p. xvii). In short, for Keith, the foundation of his case for geoengineering lies in the post-natural turn in environmental thinking.

This is accompanied by a political stance, expressed as a humanitarian desire to help the world's poorest. At the centre of what Keith imagines SGE will bring is the idea that SGE will reduce climate risk and thereby

---

[8] If one assumes SGE to be primarily a technoscientific object, a 'thing' or set of things which must be made to operate reliably to spray aerosols into the stratosphere and achieve cooling effects, then 'mundane' is an apt description. It builds upon pre-existing technologies and existing knowledge of physics and chemistry. However, if SGE is understood as a sociotechnical, world-shaping, project then it is not mundane at all.

"benefit the poor and politically disadvantaged" by reducing crop losses, heat stress and flooding over the next fifty years (Keith 2013: 10).

Keith is explicit about his differences with opponents of geoengineering. Local is not better than global: "the essence of solving global commons problems like climate is to compel local communities to cut their emissions … [against their] self interest" (Keith 2013: 146). On technology he argues that "most of the big environmental wins of the last half century have been techno-fixes," not "social fixes" (p. 147). On capitalism he argues that calls for "a fundamental reengineering of market capitalism" are misplaced: since it does a better job of environmental regulation than any other system (p. 145). And on the possibility of global inequality effects he maintains that whilst it is possible SGE "will bolster the power of the strong over the weak … the opposite seems just as likely" (p. 153). The cheapness of SGE means it is "a levelling technology" and "acts to diffuse power away from the richest and most powerful nations to a much larger pool of weaker states" (p. 155). It is "a cheap tool that could green the world" (p. x).

Couched in the language of values, risk and rationality, and with the self-assurance to proclaim SGE as being part of a new environmentalism and also good for the poor, this is the voice of power. It is the language from on high of global solutions to global problems. It is akin to the World Bank and IMF prescriptions of what is needed for the greater good, for stability, for development. Despite all the qualifications, to be solved by further research and trial deployment, SGE is offered as salvation. The choice of cover image for Keith's book is revealing (see Fig. 4.2), since it mirrors typical Christian iconography so closely. It depicts, in grayscale tones, a shaft of sunlight piercing through dark clouds onto the peaceful surface below.

Chemtrails: Secret Elites Poisoning the World

> For many decades a covert global collaboration of governments has participated in the complete manipulation/disruption of the planet's weather and climate systems. There are many agendas … [none] benevolent.—
> Geoengineering Watch (2017a)

4 Competing Imaginaries of Solar Geoengineering 141

Fig. 4.2 Book cover which encapsulate the 'Salvation' narrative (Courtesy: MIT Press)

The Chemtrail imaginary revolves around the idea that climate and weather engineering is already happening. Proponents argue that the evidence for this can be seen in the contrails often visible behind certain aircraft. These contrails aim to change the climate and they contain poisonous substances being secretly, deliberately and maliciously dumped on ordinary people. Adherents of the Chemtrail imaginary differ as to who the elites getting government to do this are: bankers (sometimes Jewish bankers is the suggestion), the Illuminati, the military and, more typically today, the 'powerful' generally. Many poisonous chemicals are said to be involved: aluminium, arsenic, barium and much else besides. The Chemtrailers' major concern is to expose and oppose this conspiracy.

Why the elites would be doing this is rarely fleshed out: perhaps to reduce population is sometimes the suggestion, but a generalized assumption that this is what "they" do pervades the websites, as does a desire to expose this conspiracy. The idea of malicious use of chemtrails predates the re-emergence of SGE in the mid-2000s and has been re-formulated and attached to SGE more recently.

Whilst there is no single Chemtrail imaginary, variants can be found on a variety of websites. Major ones, at the time of writing, include GeoengineeringWatch and GlobalSkyWatch amongst many, many others (Cairns 2014). One prominent site, Global Research, calls itself a centre for research on globalisation, itself suggestive of where it places blame for chemtrails. Whilst the sites are overwhelmingly based in the United States, significant followings and local sites can be found in a number of European and other 'developed' countries. Some of the Chemtrail advocates have paid to attend academic geoengineering conferences where they frequently film proceedings to obtain "proof" that solar geoengineering is happening, and they have attended, for example, the launch of the NRC reports on climate engineering in 2015.

The Chemtrail account focuses on SGE rather than other forms of geoengineering. It has a significant internet presence. The numbers familiar with and accepting of its account easily exceed those familiar with geoengineering as a climate policy. Indeed, a study by Tingley and Wagner found that in mid-2017 10% of Americans strongly believed that "the government has a secret program that uses airplanes to put harmful chemicals into the air [chemtrails]", and a further 29% were inclined to believe this (Tingley and Wagner 2017: 2). These figures were sharply up on surveys in previous years. Although the 2017 survey showed that belief in the Chemtrails account spanned the political spectrum, the authors argue that subsequent analysis of social media suggests the rise is associated with the Trump phenomenon and with an increase in distrust of science as an institution and with 'unwarranted fears' around issues such as the autism fears of anti-vaccination activists (Tingley and Wagner 2017: 5). It is revealing that the comments sections of these websites frequently make a connection between the chemtrails poisoning conspiracy, as they see it, and other conspiracy theories (such as the JFK assassination). Noteworthy, too, is that in 2016 website comments also frequently expressed admiration for Donald Trump. But this is not uniformly the case. As a detailed report by Dunne (2017) reveals, those convinced of the chemtrails account can include ordinary Americans who otherwise lean towards Democratic Party sympathies.

If one takes the GeoengineeringWatch website as the most articulate contemporary proponent of the Chemtrail argument then one observes the following. Climate engineering is seen as only one of a number of "highly toxic assaults" that "we are collectively being forced to endure." Others include fluoridated water, "chemical laden GMO foods" and forced vaccinations (Geoengineering Watch 2017b). Together these are impacting "our cognitive ability" and causing intelligence levels to fall. Geoengineering, rather than climate change, is understood to explain changing weather patterns, destruction of the ozone layer and "toxic fallout" generally (Geoengineering Watch 2014). The fact of climate change is no longer vigorously denied as it once was by Chemtrailers, but neither is it explicitly acknowledged. The seeds of a more environmentally-attuned stance are increasingly evident as is a move to more standard accounts of the elites deemed responsible: "To interfere with Earth's life support systems is insanity beyond comprehension and this is exactly what the completely out of control military industrial complex has done" (Geoengineering Watch 2014).

Chemtrailers typically summon people to "wake up" and "look up" and to recognise that malign government (on behalf of global elites) are engaging in a secret geoengineering programme to poison people (and their land, water and air) by spraying dangerous chemicals and biological agents from aircraft at high altitudes and claiming that these are the normal contrails produced by aircraft. The sites take great pains to claim scientific authenticity and to express their findings in a scientific idiom. For example they produce test results, usually undertaken by supporters, of soil contamination although they are less clear when showing a link between these and the aircraft contrails. More recently the GeoengineeringWatch website has said that SGE is linked to 'many agendas' and now includes a more conventional-sounding environmental account linking SGE to biosphere damage more generally:

Every engineered (and highly toxic) chemical cool-down event that the geoengineers carry out comes at the cost of an even worse overall warming of the planet. It comes at the cost of an even more damaged ozone layer and an even more contaminated biosphere. The human race has inflicted countless forms of damage to our once thriving planet (with the ongoing geoengineering insanity at the top of the list). How much more can the planet take before complete biosphere implosion? (Geoengineering Watch 2017a)

Despite their conflicting stances both the Un-Natural and Imperial imaginaries are in some sense part of mainstream discourse surrounding SGE and climate change. However, the Chemtrail imaginary operates in a different register. The former are familiar with the dynamics of climate change, both attempt to marshal reputable evidence in support of their arguments (although they may interpret it variously), and both can be found to a greater or lesser extent in the major institutional assessments of SGE. The Chemtrail imaginary, by contrast, lurks in the shadows of mainstream climate discourse. It contains diverse suppositions about the "reality" of climate change, compared to the other two imaginaries. Some Chemtrail websites make no connection to climate change. Others see climate change as another elite hoax. Yet others suggest that the spraying was first done to change climate and is now done to cover up that "fact". It is very different too in assuming that SGE is currently happening and has been happening for a long time. Increasingly, the chemtrail sites talk about solar geoengineering rather than chemtrails. The Chemtrail imaginary makes extensive use of the language of science but the evidence and findings it relies upon are not typically recognized by mainstream climate science.

Some mainstream scientists have replicated the strategy adopted in relation to climate change scepticism/denialism and have produced papers showing that 99% of scientific experts agree that a secret large-scale atmospheric spraying programme (SLAP) does not exist (Shearer et al. 2016). Others have set up websites which specialise in debunking chemtrail ideas as pseudo-science, notably ContrailScience, in the belief that perceived consensus is a "key 'gateway' cognition that acts as an important determinant of public opinion" (Maibach and Van der Linden 2016). But it could equally be argued that discrediting the Chemtrail imaginary as mere conspiracy theory, or a manifestation of the "post-truth" turn in politics, or as unsupported by facts or experts, is both unlikely to reduce the number of its adherents and also to miss the point. The Chemtrail imaginary is reflective of a deep suspicion about the motives and actions of the powerful, especially in the context of market globalism. In this it has much in common with the Un-Natural imaginary.

The significance of the Chemtrail imaginary lies less in the facticity of its specific claims than in the disenchantment with expert claims and the suspicion of elite initiatives, especially transnational ones, that it reveals. The size of its following especially in the United States, the country most likely to endorse or deploy SGE, means that its impact on how SGE is understood and whether it is regarded as politically acceptable should not be underestimated.

Convergences and Divergences in the Imaginaries

Table 4.1 summarizes schematically the key elements of the three competing imaginaries discussed above with the italics areas indicating elements visibly present in major institutional assessment reports. This includes not only the general world outlook which animates each and their general stance towards the dominant global order, but also a range of elements of specific relevance to SGE: such as their understanding of the contemporary climate condition, of 'nature', as well as of technology. Mood too is important: hegemonic narratives generally project a vision of a desirable future social order associated with the technology. Optimism has, to date, proven especially challenging for all but the most blindly enthusiastic proponents of SGE.

None of these imaginaries is hegemonic although elements of some are more apparent than others in the major institutional assessments of SGE. Whilst each imaginary is distinct, they often share assumptions and perspectives.

The Market, Geo-management and Salvation narratives which make up the Imperial imaginary of SGE have different points of emphasis and approach the issue from different disciplinary standpoints (economic, geo-political and eco-modernist respectively). And yet all broadly imagine

Table 4.1 Summary of competing imaginaries of solar geoengineering

	Un-Natural	Market	Imperial Geo-management	Salvation	Chemtrail
SGE seen as …	Dangerous elite initiative which must be opposed	Opportunity for capitalist-friendly climate solution	*Essential addition to portfolio needed to tackle climate change*	Essential to save world's poorest from dangerous climate change	Elite conspiracy to poison people and planet
World outlook	Global North dominates the world and resists both strong climate action and justice for the South RADICAL (SOUTHERN) ALTERITY	Economically neo-liberal. Most problems solvable by removing barriers to free enterprise MARKETS, GROWTH & DEVELOPMENT	Contain climate catastrophe by getting 'countries that matter' to take realistic action REALIST POLICIES + TECHNOLOGY	*Avoid climate catastrophe and make the world safe for modernity and development* THERE IS NO ALTERNATIVE	Elites are conspiring against ordinary (1st world) citizens to poison people and pollute the planet RESIST ELITE CONSPIRACY
Capitalism/ dominant order	Big capital causing climate problem; dominance of global North a problem	Strongly favor capitalism; Assumes US geo-political hegemony *Rational calculation (utilitarian cost-benefit) model key to decision*	*Existing economic and geo-political order are the 'reality'*	Flawed but adaptable; no need for fundamental change	Capitalism not challenged but suspicious of monopoly and finance capital/ bankers and military-industrial complex

(continued)

Table 4.1 (continued)

	Un-Natural	Market	Geo-management	Salvation	Chemtrail
			Imperial		
Climate Change	A major problem with existing answers (mitigation, climate justice, equitable carbon budgets, CBDR, etc.)	Potential future risk/problem. Existing policies (mitigation, pricing carbon) expensive and growth-reducing	A major problem, but an energy markets and technology problem not an environmental one	A major problem where mitigation is too slow and unlikely to be enough.	If it exists, global warming is caused by previous and current military and geoengineering interventions
Nature/ Environment	Avoid unnecessary, deliberate + un-natural intervention into complex Earth systems (Mother Earth/Gaia)	Nature secondary to human interests and economic imperatives.	Manage the environment by adapting to problems or masking them	Post-nature: hybrid environments shaped by people	Geoengineering involves poisoning people and the environment
Democracy/ Justice	Undemocratic and unjust if climate response decisions made in the global North	Democracy may constrain action to develop SGE	Club of powerful states must co-operate (not all states). Silent on justice. Democracy subservient	Justice (as Development) important motivation for SGE. Silent on democracy	*Largely silent on democracy*
Science & Technology	Need localized, humble tech not globalised, big science technofixes; Uncertainties and dangers too great	Embrace technology and technofixes	Affordable, effective technologies the key answer to environmental problems; Brave New World	*Technofixes where needed.* SGE more engineering than science; resolve uncertainties in practice	Uses scientific language but rejects expert claims; hostile to some technologies

(continued)

Table 4.1 (continued)

	Un-Natural	Imperial			Chemtrail
		Market	Geo-management	Salvation	
Mood	Dystopian present and chaos ahead if SGE imposed on the world	*Bland, matter-of-fact and technical*	Miserabilist Realism	*Distasteful option but 'lesser evil'*	Dark apocalyptic outlook
Geoengineering implication	Risky and dangerous. Ban deployment; constrain research; resist imposition on global South	Actively develop and deploy if not too obviously risky	Get it ready as it may be needed, even if risky	Develop, incrementally deploy, monitor, further develop etc... *Use as part of portfolio of action (including mitigation).*	Need to expose current geoengineering and stop it

Note: italics sections means some presence in institutional reports and assessments a continuation of the dominant global order, what Steger has termed "market globalism" (Steger 2009) and presume a leading and largely benevolent role for the United States. And they all share a belief that new and large-scale technology is the most critical component when it comes to addressing climate change, even embracing explicitly the desirability or inevitability of 'technofixes.' The Un-Natural imaginary, by contrast, uses this term disparagingly, is skeptical of big technology and big science, and is also explicitly critical of the dominant global order. It shares with the Chemtrail imaginary a suspicion of powerful elites. Both the Un-Natural imaginary and the Salvation version of the Imperial imaginary are alert to, and troubled by, global inequality and poverty, although they differ sharply as to the best response to this: with the former imagining 'bottom-up' solutions (*with* 'the poor'), and the latter imagining 'top-down' ones (*for* 'the poor').

'Dangerous climate change' is a regular trope in mainstream accounts of climate change (Liverman 2009). The Un-Natural and the Imperial imaginaries generally take the climate prognosis to be extremely serious and something to be addressed urgently. A discourse of climate 'crisis' permeates their imaginings of SGE although each processes this through different interpretative lenses: as a crisis of global injustice (Un-Natural), a crisis of policy and global order (especially in the Geo-management narrative), and, more familiarly, as a crisis of rising temperatures (Un-Natural and Imperial). It is certainly true that in the Market narrative presented here the urgency and immediacy of the issue is downplayed, even as the 'long tails' of climate risk are acknowledged. However, as the Richard Branson quotation cited earlier suggests, there are versions of the Imperial imaginary's Market narrative which are more in line with mainstream views of the climate situation as both serious and urgent. Both the Un-Natural and the Imperial imaginaries embrace a view of climate as a 'global' object, although the Un-Natural imaginary is the only one to stress that climate change is driven and experienced differentially and lived locally, and should be responded to accordingly. The Chemtrail imaginary is the outlier here as it either dismisses or, more recently, avoids engaging with the standard understandings of climate change.

The dominant approach to climate change, as found in the work of the IPCC, privileges a scientific framing of the issue and a utilitarian analysis of necessary responses (O'Lear 2016; Liverman 2009). This is also the approach adopted in the main institutional assessments of geoengineering. Intriguingly, although both the Un-Natural and Imperial imaginaries have an adequate grasp of climate science, none of them prioritizes a purely science-based framing of SGE, in the sense of suggesting that policy flows from the science. Policy is understood to be a choice, and the world of SGE that must be imagined needs to be underpinned by more recognizably social (and economic) concerns. Both imaginaries focus on evidence supportive of their arguments and either ignore or attempt to refute evidence which does not. The Un-Natural imaginary, perhaps paradoxically, comes closest to an approach centered in the scientific literature. It places heavy reliance in its argument against SGE on the general insight from the Earth Sciences that Earth systems are complex and inter-linked, with deeply uncertain dynamics, and should therefore not be tampered with. This embrace of science is mobilized to suggest any interference with such complex natural systems should be avoided.

The many surveys of public opinion regarding SGE report that much of the discomfort with the idea of geoengineering is because people find what it is proposed to do to the Earth to be in some sense 'un-natural' (Bellamy and Lezaun 2017). The Un-Natural and the Chemtrail imaginaries both pay attention to the question of nature. For the Chemtrailers SGE entails poisoning the earth and its people. In the Un-Natural imaginary SGE is imagined as a violation of Gaia and of natural laws, an offence against 'Mother Earth', or Pachamama, and 'Father Sky'. Concern for nature is less evident in the Imperial imaginary, although this is not as true for the Salvation strand of that imaginary. We have seen that Keith regards himself as an environmentalist and a dedicated lover of wilderness, and is clearly troubled by the Nature problem. He lands on the position that there is no longer any true wilderness, that we live in hybrid environments, and that the question therefore becomes what elements of the non-human we choose to value and, accordingly, how we shape the world. This is post-nature thinking in the Breakthrough Institute tradition, and indeed, Keith is a founder signatory of their *Ecomodernist Manifesto* (2015). SGE becomes, therefore, a part of that hybrid 'made' world, something to be managed. It may have its monstrous side, but if so, as Latour has implied, it is a monster to be loved and for which we must assume responsibility, rather than one to be abandoned (2012). In the Geo-management version of the Imperial imaginary 'nature' is barely discussed, other than as something to be managed. In the Market version nature is ignored as an issue, apart from a few disparaging comments about environmentalists, and it is assumed that nature is secondary to human interests and economic imperatives. To the extent that the three imaginaries pay attention to the question of nature, they adopt different standpoints in relation to its value, its distinctiveness from the human, and its subjective centrality.

The Chemtrail and Un-Natural imaginaries share some points in common but they also differ. The Un-Natural imaginary, unlike the Chemtrail one, is rooted in mainstream understandings of climate change and climate science. It also accepts that SGE is not yet happening, that it is a proposed intervention not a long-extant one. But the Chemtrail and Un-Natural imaginaries share a sense of being outside the corridors of power, although whilst the former is anti-big government, the latter is anti- the dominant order. Both have a conspiratorial edge in their accounts of SGE—with the Chemtrailers seeing the hand of secret cabals at work, whilst the Un-Natural imaginary adopts a more structural account of military, big science and big money collaborating (ETC/Biofuelwatch 2017: 42–3). Both are suspicious, in different ways, of the workings of global elites and of big technologies in the hands of those elites. In this they are distinct from the Imperial imaginary and its sense of being close to mainstream expressions of economic power, United States power, and liberal scientific power respectively (emphasized respectively in the Market, Geo-management and Salvation versions). Whilst the Imperial imaginary speaks to policymakers and to power, the other two imaginaries speak against power.

A Sociotechnical Imaginary Struggling to Be Born

Any prospective technology emerges concurrently with a story, typically optimistic, of how it might fit into and even shape an envisioned future: think of nuclear power imagined at its inception as inaugurating a world with "electricity too cheap to meter". Such imaginaries are animated by particular understandings of how social order is, and should be, structured in the world. They contain and reproduce worldviews, including how the world is ordered, how power is enacted, and to what ends. They are rooted in the present even as they envision particular futures. According to Jasanoff, the concept of the 'sociotechnical imaginary' refers to "collectively held, institutionally stabilized, and publicly performed visions of desirable futures, animated by shared understandings of forms of social life and social order attainable through, and supportive of, advances in science and technology" (Jasanoff 2015: 3).

The three strands of the Imperial imaginary I have outlined here—Market, Salvation and Geo-management—all imagine a continuity and compatibility with the currently dominant global political and economic order. All are focused on getting SGE taken seriously as a climate policy option and enabling its development and trialing to proceed. They each map onto a different part of the internal landscape of US politics. The Salvation narrative is closer in outlook to the Democratic establishment, and the Market narrative to the Republican right, whilst the Geo-management perspective is linked to often bi-partisan perspectives about exercising US power in the world. But despite these different nuances they are entwined narratives. This is evident, for example, in the frequency of reference to SGE's cheapness in both the Geo-management and Salvation narratives and the framing of the climate problem in the language of risk-risk trade-offs.[9]

Indeed the three strands are coalescing, I argue, into an Imperial sociotechnical imaginary of SGE. This imagines SGE as a powerful technology, which those with global power (and only those) should be able to deploy, without disrupting the existing systems and centers of global economic and political power, and deployed in the purported interests of the least powerful. This Imperial imaginary has been given an extensive hearing in the major institutional assessments of SGE.[10] And

---

[9] The version of the Market narrative I have selected as an exemplar is especially ideologically partisan. This may explain why the Salvation narrative is often embarrassed by the implications of embracing such strongly utilitarian principles when it comes to the problem of climate change, whilst still acknowledging the cheapness of SGE and much of the cost-benefit modelling which flows from the Market narrative.

[10] In practice, David Keith and David Victor have both been particularly influential in shaping the institutional reports (as panelists and reviewers), and many other key panelists and the small circle of SGE knowledge-brokers appear to broadly share their vision. Proponents of the most extreme versions of the Market narrative have also been represented although to a slightly lesser extent.

4 Competing Imaginaries of Solar Geoengineering 153

yet it has not managed to become hegemonic. Its proponents have, to date, been unable to persuade the majority of climate policymakers and climate scientists that SGE should be adopted as an additional leg of climate policy. They have failed to generate a more stable and widely embraced vision of why SGE is needed and how it should proceed. If anything, between the Royal Society report (2009) and the NRC report (2015a) their influence may have diminished. Understanding why this the case is important both for proponents of SGE and for those hostile to the geoengineering turn in climate policy, even if, as I will explore later, that influence may again be on the rise following the Paris agreement.

In her work on sociotechnical imaginaries, Jasanoff notes that they "can originate in the visions of single individuals, gaining traction through blatant exercises of power or sustained acts of coalition building" (Jasanoff 2015: 3). In the case of SGE it has not been possible to simply impose a vision of SGE and enact it. Efforts to include it as part of US climate policy, most recently under the George W. Bush administration, have been unsuccessful, finding few champions. Whilst the Trump administration includes some enthusiasts for SGE, these proponents typically do not regard climate change as a priority issue. Coalition-building has therefore been the strategy adopted. There have been, and continue to be, sustained and well-financed efforts aimed at building broadly supportive coalitions and finding allies outside the United States. David Keith has been especially influential in efforts both to normalize SGE as a component of climate policy and to give a green light to accelerating research into and trial deployment of SGE. The nascent Imperial imaginary is now shared by a significant and influential, but still relatively small, network of 'knowledge brokers' mainly located in the global North. But this imaginary has failed to gain more generalized traction amongst climate and policy experts. Even a cursory reading of the IPCC, NRC and SRMGI reports indicates that sceptical or lukewarm attitudes to SGE prevail (IPCC 2012; NRC 2015a, b, c, d; SRMGI 2011). At the extreme, one panel member for the NRC report, the eminent scientist Raymond Pierrehumbert, was moved to state: "The nearly two years' worth of reading and animated discussions that went into this study have convinced me more than ever that the idea of 'fixing' the climate by hacking the Earth's reflection of sunlight is wildly, utterly, howlingly barking mad" (2015).

The Imperial Imaginary's Quest for Legitimacy

The Imperial imaginary of SGE has been unable to rise to dominance for policy purposes. Notwithstanding the influence of its key proponents and their discursive proximity to the established order they have, to date, been unable to mobilize sufficient political power or discursive consensus to impose their vision of SGE or gain widespread support from climate policymakers. By contrast, the sociotechnical imaginaries manifest in both the Chemtrail and the Un-Natural imaginaries have, in different ways, proven to be remarkably resonant with public opinion, and something like the Un-Natural imaginary is commonly found amongst those concerned with climate change and climate policy. This is notable especially given the lack of resources behind these views and their oppositional stance towards established power. How does one explain why the Imperial sociotechnical imaginary of SGE has been unable to become hegemonic? This is clearly constraining SGE from being 'officially' endorsed and embraced as a respectable and enactable component of climate policy. In addressing this question my focus for now is on the ideational and the imagined. Physical climate developments will be considered in later Chapters.

One explanation can be found in the imbalance between the utopian and the dystopian. Jasanoff has noted that dominant sociotechnical imaginaries are "typically grounded in positive visions of social progress" even when they acknowledge a shadow side (Jasanoff 2015: 3). In the case of SGE none of the imaginaries outlined above is able to present a positive vision. Indeed, the dystopian overwhelms even the suggestion of utopian. Buck (2010) has accurately noted that most narratives of geoengineering are stories of failure and insurance. Keith's Salvation narrative comes closest to a positive vision in imagining SGE as a technology to help save the poor from serious climate change. But even he primarily presents SGE as regrettable, risky, distasteful and, at best, a necessary 'lesser evil.'

Another explanation may lie in the strength, persistence and public resonance of the idea that SGE is un-natural and therefore unwise. The claim of SGE's necessity struggles to break through this centuries-old, ontological barrier between the social and the natural. Whilst the taboo against investigating SGE may have been broken in the specialist community of climate science and climate policy, it has not been broken more broadly. Indeed, the persistence of the largely un-resourced Un-Natural narrative and the spread of the Chemtrail narrative suggests a deep-rooted, visceral hostility to SGE as both un-natural and an elite imposition. My own experience, when presenting the idea of SGE in the most neutral manner I can muster, has been that audiences are typically shocked to find the idea is even under consideration, and to wonder why scientists assume they can 'play God'.

Further, as I have shown in Chapter 2, there are substantial indications, especially since the 1970s and 1980s, of a change in the normative stance of many of the relevant scientific experts, such as climate scientists and Earth scientists. A significant proportion of these experts now display an attachment to ideas of 'nature' having inherent value, to a sense that limits/boundaries/thresholds carry normative implications for social practice, and that ideas such as 'sustainability' imply some need for retreat from existing consumption practices with their attendant ecological footprint. Pushing back against such attitudes is what Victor perhaps has in mind when he charges environmental scholars with "car[ing] too much about their subject" and implies they are too caught up in "green evangelism" (2011: x–xi). It helps explain the persistence of the idea of SGE as 'un-natural', even as that narrative sits outside the circles of power considering SGE as a policy alternative. It also explains Keith's efforts to summon concepts such as 'post-nature' and hybridity in order to present a vision which can cut through talk of anything being un-natural.

Yet another explanation for the failure of the Imperial imaginary to become dominant may lie in its incomplete engagement with questions of material and geo-political power. Is a world of SGE safer (and for whom?) than a world without it? Clearly there is interest in SGE in both the US security establishment (prime sponsors of the NRC report) and amongst at least parts of the existing elite. The attraction of SGE for the powerful lies in the promise of a technological intervention which can be climatically stabilizing, which can reduce climate risk relatively cheaply and without seriously disrupting the established global economic and military order. But SGE can also, credibly, be expected to be deeply de-stabilizing of the geo-political order, as Healey and Rayner (2015) have argued using the example of India and Pakistan. In coming to terms with these issues, much depends upon who it is imagined will make deployment decisions (raising questions of democracy, justice and governance) and what the climate and agricultural effects of intervention actually turn out to be (currently expectations in this regard are marked by significant uncertainty). The security agencies of the global superpower are also anxious about 'rogue' deployment (deployment which it has not authorized) and possibly resulting in an SGE 'arms race'. Whilst the focus of the Geo-management narrative is on questions such as these, the vision it offers of a 'club' of powerful states ushering in a Brave New World is not only un-democratic, techno-centric and gloomy. More important, for the established power elites, it is far from clear that it would make the world safer for them.[11]

In short, I suggest that pessimism/dystopia, insufficient acknowledgement of traditional views that SGE is un-natural and unwise, the normative stance of many within the field of climate science and policy, and a failure to produce a convincing account that SGE would make the world safer for established elites, all help explain why SGE, and especially the Imperial vision of it, has not become broadly accepted. Or perhaps it is simply the case that a sociotechnical imaginary must always be contested when the technology itself mainly exists theoretically and is not yet deployed. In any event, the reality remains that what SGE is, what it is for, and whether it is desirable, are all disputed today. No single imaginary is dominant. SGE is best understood as a technology and a sociotechnical imaginary struggling to be born. Not surprisingly, contestation about power, knowledge and values pervades contemporary research into SGE. It is to this that I now turn.

References

Bellamy, R., & Lezaun, J. (2017). Crafting a public for geoengineering. *Public Understanding of Science, 26*(4), 402–417.

Bickel, J. E., & Agrawal, S. (2013). Reexamining the economics of aerosol geoengineering. *Climatic Change, 119*(3–4), 993–1006.

---

[11] When the Geo-management and Salvation narratives are attached, a 'liberal' version (in US political terminology) might emerge—the Brave New World is needed to save the poor. When the Geo-management narrative is attached to the Market narrative a more isolationist, security-focused version emerges—the global North (plus China?) fences itself off, uses SGE and adapts to climate change, whilst leaving the Rest to their fates.

Bickel, J. E., & Lane, L. (2012). *Challenge paper: Climate change, climate engineering R&D*. Copenhagen: Copenhagen Consensus Center.
Bipartisan Policy Center (BPC). (2011). *Geoengineering: A national strategic plan for research on the potential effectiveness, feasibility, and consequences of climate remediation technologies*. Washington, DC: Bipartisan Policy Center Task Force on Climate Remediation Research.
Blackstock, J. J., Battisti, D. S., Caldeira, K., Eardley, D. M., Katz, J. I., Keith, D. W. et al. (2009). *Climate engineering responses to climate emergencies*. Santa Barbara: Novim. Retrieved January 9, 2019, from http://arxiv.org/pdf/0907.5140
Bloor, D. (1991). *Knowledge and social imagery* (2nd ed.). London: University of Chicago Press.
Buck, H. J. (2010). What can geoengineering do for us? Public participation and the new media landscape. *Unpublished workshop paper*. Retrieved January 9, 2019, from http://www.umt.edu/ethics/ethicsgeoengineering/Workshop/articles1/Holly%20Buck.pdf
C2G2 (Carnegie Climate Geoengineering Governance Initiative). (2018, November). *Governing Solar Radiation Modification (SRM)*. Retrieved January 9, 2019, from https://www.c2g2.net/wp-content/uploads/C2G2_Solar-Brief-hyperlink.pdf
Cairns, R. (2014). Climates of suspicion: 'Chemtrail' conspiracy narratives and the international politics of geoengineering. *Climate Geoengineering Governance Working Paper Series*: 009. Retrieved January 9, 2019, from http://geoengineering-governance-research.org/perch/resources/workingpaper9cairnsclimatesofsuspicion-1.pdf
Dunne, C. (2017, May 22). My month with chemtrails conspiracy theorists. *The Guardian*. Retrieved January 9, 2019, from https://www.theguardian.com/environment/2017/may/22/california-conspiracy-theorist-farmers-chemtrails
Ebell, M. (2009). The anti-green ecologist. Retrieved January 9, 2019, from https://cei.org/op-eds-and-articles/anti-green-ecologist
Ecomodernist Manifesto. (2015). Retrieved January 9, 2019, from http://www.ecomodernism.org/
Ellison, K. (2018, March 28). Why climate change skeptics are backing geoengineering. *Wired*. Retrieved January 9, 2019, from https://www.wired.com/story/why-climate-change-skeptics-are-backing-geoengineering/
ETC Group. (2009). *Retooling the planet? Climate chaos in the geoengineering age*. Commissioned by the Swedish Society for Nature Conservation, prepared by ETC Group. Stockholm: Swedish Society for Nature Conservation. Retrieved

April 20, 2019, from http://www.etcgroup.org/sites/www.etcgroup.org/files/publication/pdf_file/Retooling%20the%20Planet%201.2.pdf

ETC Group. (2010a, November). *Geopiracy: The case against geoengineering* (2nd ed.). Retrieved January 9, 2019, from https://www.cbd.int/doc/emerging-issues/etcgroup-geopiracy-2011-013-en.pdf

ETC Group. (2010b). *Geoengineering: Gambling with Gaia*. ETC Group Briefing Paper ahead of COP10, October 2010. Retrieved September 22, 2011, from http://www.etcgroup.org/upload/publication/pdf_file/ETC_COP10GeoBriefing081010.pdf

ETC Group. (2013a). The artificial intelligence of geoengineering. Communique #109, Jan/Feb. Retrieved April 20, 2019, from http://www.etcgroup.org/content/artificial-intelligence-geoengineering

ETC Group. (2013b). Normalizing geoengineering as foreign aid. The Artificial Intelligence of Geoengineering Part 3, Blog post 4 April 2013. Retrieved April 20, 2019, from http://www.etcgroup.org/content/normalizing-geoengineering-foreign-aid

ETC Group. (n.d.). Mission and current focus. Retrieved January 9, 2019, from http://www.etcgroup.org/mission

ETC Group/Biofuelwatch. (2017). *The big bad fix: The case against climate geoengineering*. Retrieved January 9, 2019, from http://etcgroup.org/sites/www.etcgroup.org/files/files/etc_bbf_mar2018_us_v1_web.pdf

Fleming, J. R. (2010). *Fixing the sky: The checkered history of weather and climate control*. New York: Columbia University Press.

GAO (Government Accountability Office). (2010). *Climate change: A coordinated strategy could focus federal geoengineering research and inform governance efforts*. GAO-10-903. Washington, DC: U.S. Government Accountability Office. Retrieved January 9, 2019, from https://www.gao.gov/assets/320/310105.pdf

Geoengineering Watch. (2014, March 2). Climate engineering introduction letter. Retrieved January 9, 2019, from https://www.geoengineeringwatch.org/flaming-arrow-package/

Geoengineering Watch. (2017a, May 2). Climate engineering completely manipulating precipitation. Retrieved January 9, 2019, from https://www.geoengineeringwatch.org/climate-engineering-completely-manipulating-precipitation/

Geoengineering Watch. (2017b, May 20). Geoengineering Watch global alert news. Retrieved January 9, 2019, from https://www.geoengineeringwatch.org/geoengineering-watch-global-alert-news-may-20-2017/

Gingrich, N. (2008, June 3). Stop the green pig: Defeat the Boxer-Warner-Lieberman green pork bill capping American jobs and trading America's future. *Human Events*. Retrieved January 9, 2019, from http://humanevents.com/2008/06/03/stop-the-green-pig-defeat-the-boxerwarnerlieberman-green-pork-bill-capping-american-jobs-and-trading-americas-future/

Goes, M., Tuana, N., & Keller, K. (2011). The economics (or lack thereof) of aerosol geoengineering. *Climatic Change, 109*(3–4), 719–744.

Healey, P., & Rayner, S. (2015). Key findings from the Climate Geoengineering Governance (CGG) project. *Climate Geoengineering Governance Working Paper Series*: 25. Retrieved January 9, 2019, from http://www.geoengineering-governance-research.org/perch/resources/workingpaper25healeyraynerkeyfindings-1.pdf

HOME (Hands Off Mother Earth). (2018) *Manifesto against geoengineering*. Retrieved January 9, 2019, from http://www.etcgroup.org/sites/www.etcgroup.org/files/files/home_manifesto_english_.pdf

Hulme, M. (2014). *Can science fix climate change? A case against climate engineering*. Cambridge: Polity Press.

IPCC (Intergovernmental Panel on Climate Change). (2012). *Meeting report of the Intergovernmental Panel on Climate Change expert meeting on geoengineering* (O. Edenhofer, R. Pichs-Madruga, Y. Sokona, C. Field, V. Barros, T. F. Stocker, Q. Dahe, J, Minx, K. Mach, G.-K. Plattner, S. Schlömer, G. Hansen, & M. Mastrandrea, Eds.). IPCC Working Group III Technical Support Unit, Potsdam Institute for Climate Impact Research. Geneva: IPCC.

Jasanoff, S. (2015). Future imperfect: Science, technology and the imaginations of modernity. In S. Jasanoff & S.-H. Kim (Eds.), *Dreamscapes of modernity: Sociotechnical imaginaries and the fabrication of power* (pp. 1–33). Chicago: Chicago University Press.

Keith, D. W. (2013). *A case for climate engineering*. Cambridge, MA: MIT Press.

Lane, L., & Bickel, J. E. (2013). *Solar radiation management: An evolving climate policy option*. Washington, DC: American Enterprise Institute.

Lane, L., Caldeira, K., Chatfield, R., Langhoff, S. (2007). *Workshop report on managing solar radiation*. NASA Ames Research Centre & Carnegie Institute of Washington, Moffett Field, CA, 18–19 November. Hanover, MD: NASA. (NASA/CP-2007-214558).

Latour, B. (2012). Love your monsters. *Breakthrough Journal*, 2. Retrieved January 20, 2019, from http://thebreakthrough.org/index.php/journal/past-issues/issue-2/love-your-monsters

Liverman, D. M. (2009). Conventions of climate change: Constructions of danger and the dispossession of the atmosphere. *Journal of Historical Geography, 35*, 279–296.

Maibach, E. W., & Van der Linden, S. L. (2016). The importance of assessing and communicating scientific consensus. *Environmental Research Letters, 11*(9), 091003.

National Research Council (NRC). (2015a). *Climate intervention: Reflecting sunlight to cool Earth.* Washington, DC: National Academy of Sciences.

National Research Council (NRC). (2015b). *Climate intervention: Summary report.* Washington, DC: National Academy of Sciences.

National Research Council (NRC). (2015c). *Climate intervention: Carbon dioxide removal and reliable sequestration; reflecting sunlight to cool Earth.* Washington, DC: National Academy of Sciences.

National Research Council (NRC). (2015d). *Climate intervention reports release briefing webcast.* Retrieved January 9, 2019, from http://nas-sites.org/americasclimatechoices/videos-multimedia/climate-intervention-reports-release-briefing-webcast/

Nordhaus, W. (2007). *The challenge of global warming: Economic models and environmental policy.* Retrieved January 9, 2019, from http://www.econ.yale.edu/~nordhaus/homepage/OldWebFiles/DICEGAMS/dice_mss_072407_all.pdf

O'Lear, S. (2016). Climate science and slow violence: A view from political geography and STS on mobilizing technoscientific ontologies of climate change. *Political Geography, 52*, 4–13.

Olson, R. L. (2011). *Geoengineering for decision makers.* Science and Technology Innovation Program. Washington, DC: Woodrow Wilson International Center for Scholars.

Pelkmans, M., & Machold, R. (2011). Conspiracy theories and their truth trajectories. *Focaal—Journal of Global and Historical Anthropology, 59*, 66–80.

Pierrehumbert, R. (2015, February 10). Climate hacking is barking mad. *Slate.* Retrieved January 9, 2019, from http://www.slate.com/articles/health_and_science/science/2015/02/nrc_geoengineering_report_climate_hacking_is_dangerous_and_barking_mad.html

Revkin, A. (2009, October 15). Branson on the power of biofuels and elders. *New York Times Dot Earth blog.* Retrieved January 9, 2019, from http://dotearth.blogs.nytimes.com/2009/10/15/branson-on-space-climate-biofuel-elders

Robock, A. (2008). 20 reasons why geoengineering may be a bad idea. *Bulletin of the Atomic Scientists, 64*(2), 14–18.

Royal Society. (2009). *Geoengineering the climate: Science, governance and uncertainty.* RS Policy document 10/09. London: Royal Society. Retrieved January 9, 2019, from https://royalsociety.org/~/media/Royal_Society_Content/policy/publications/2009/8693.pdf

Schelling, T. C. (2009). *Talk on geoengineering to AEI.* Retrieved January 9, 2019, from https://www.aei.org/multimedia/governing-geoengineering/

Schnare, D. W. (n.d.). *Talk on geoengineering to Heartland Institute conference.* Retrieved January 9, 2019, from http://climateconferences.heartland.org/david-schnare-iccc-1/

Shearer, C., West, M., Caldeira, K., & Davis, S. J. (2016). Quantifying expert consensus against the existence of a secret, large-scale atmospheric spraying program. *Environmental Research Letters, 11*, 084011.

Solar Radiation Management Governance Initiative (SRMGI). (2011). *Solar radiation management: The governance of research.* Issued by the Environmental Defense Fund, The Royal Society, and The World Academy of Sciences.

Steger, M. B. (2009). *Globalisms: The great ideological struggle of the twenty-first century* (3rd ed.). Lanham, MD: Rowman & Littlefield.

Stilgoe, J. (2015). *Experiment earth: Responsible innovation in geoengineering.* Abingdon: Routledge.

Tingley, D., & Wagner, G. (2017). Solar geoengineering and the chemtrails conspiracy on social media. *Palgrave Communications, 3*(12), 1–7.

US House of Representatives Committee on Science and Technology (USHCST). (2010). *Engineering the climate: Research needs and strategies for international coordination.* Report by Bart Gordon. Retrieved January 20, 2019, from https://science.house.gov/sites/democrats.science.house.gov/files/10-29%20Chairman%20Gordon%20Climate%20Engineering%20report%20-%20FINAL.pdf

Victor, D. G. (2011). *Global Warming Gridlock: Creating more effective strategies for protecting the planet.* Cambridge: Cambridge University Press.

Wagner, G., & Weitzman, M. L. (2012, October 24). Playing God. *Foreign Policy.* Retrieved January 9, 2019, from http://foreignpolicy.com/2012/10/24/playing-god/

Wigley, T. M. L. (2006). A combined mitigation/geoengineering approach to climate stabilization. *Science, 314*, 452–454.

5

Knowledge-Power-Values

In this Chapter I take a deeper dive into solar geoengineering (SGE) and explore what assumptions about power, what values and what knowledge claims accompany SGE, and how these shape and enable (or constrain) SGE's re-emergence? I explore some telling examples, focusing on the key institutional assessments and reports and the associated policy-influential literature. I will argue that these typically, even those reports which do not endorse or embrace SGE, work to normalise it as potential climate policy, facilitate its further development, and legitimise its potential deployment. They prioritise SGE's analysis as a technoscientific object (rather than a sociotechnical choice), they mask or ignore its relationship to established power, and they adopt an especially narrow reading of the ethical and values issues which SGE raises. They shore up the Imperial imaginary of SGE.

The 'is' (what the world is held to be and how this is known) and the 'ought' (what world is imagined as desirable) combine to co-produce this emergent technology and must be considered together (Jasanoff 2004). The knowledge and facts which count are central in the emergence of new technologies. They are heavily shaped by the epistemological assumptions and practices which are conscripted and the disciplinary lens which is

© The Author(s) 2019
J. Baskin, *Geoengineering, the Anthropocene and the End of Nature*,
https://doi.org/10.1007/978-3-030-17359-3_5

adopted. Imaginaries of SGE are also entwined with power and values. Not only do they have power effects but they are understood and evolve in relation to the existing global ordering, and are championed (or not) by powerful institutions, individuals, and regimes. At the risk of oversimplifying: knowledge, values and their accompanying discourses have power effects, and power shapes both what is valued, what facts matter, and indeed the structures of knowing.

In considering *Knowledge/s* I explore the elevation of scientific and engineering expertise in considerations of geoengineering, the treatment of uncertainty and the unknown, and the reliance on the calculative and utilitarian logic of cost-benefit analysis. In relation to *Power*, I explore some of the implications of SGE for capitalism and the dominant geopolitical order, as well as the ways in which the notions of emergency and risk are mobilised. Finally, in delving into *Values* I examine the ways in which the many and profound ethical questions which geoengineering raises are acknowledged but also narrowed and marginalised.

Knowledge/s

In this section I will look at three instances of how knowledge about SGE is developed. I first look at the elevation of technoscientific knowledge and paradigms in the analysis of SGE. I then look at the ways in which the many uncertainties about SGE and its possible impacts are handled, before examining the reliance on narrowly utilitarian approaches when engaging with the economics of SGE.

Epistemic Hierarchies

It is immediately apparent, on reading any of the 'institutional' reports on geoengineering, that some forms of expertise are assumed to be primary, whilst others are secondary. The reports all privilege particular forms of scientific knowledge and expertise: involving disciplines such as chemistry, atmospheric climate science, physics, climate modelling, vulcanology and similar. Only slightly down the hierarchy are engineering and economics and law. Further down still are the various social sciences and cognate disciplines (such as ethics), including efforts to gauge public opinion through surveys, focus groups and perception studies. This privileging takes place despite the widespread acknowledgement, in almost every report, and by even the most technically focussed researchers, of the significance of the ethical, the geo-political and the social in any consideration of geoengineering.

At the most obvious level one can point to the leading role taken by scientists in almost all the institutional reports and assessments. The working group responsible for the first major report, by the Royal Society (2009), only three years after Crutzen's taboo-lifting intervention, included an international lawyer, and two science and technology specialists in its otherwise science-heavy twelve-person team. The 2011 IPCC Expert Meeting on Geoengineering was likewise overwhelmingly a meeting of physical scientists, with a handful of governance specialists included (IPCC 2012: iii, 97–8). The NRC report was similarly led by physical scientists, but was slightly more inclusive. Among the non-scientists on both the committee and the review panel were experts with overtly sceptical, critical, and historical perspectives on climate engineering (2015a: v). A science-heavy focus can similarly be found in other reports, with an occasional gesture to other disciplinary fields (see for example BPC 2011; GAO 2010, 2011; Lane et al. 2007).[1]

The inclusion of non-scientists, combined with the ethical disquiet regarding geoengineering felt by some of the participating scientists, has meant that the social is not ignored. For example, whilst the focus and 'feel' of the Royal Society report is physical-science heavy, attention is drawn to the need to take account of both "the technical *and political* reversibility of each [geoengineering] proposal" (2009: 7, my emphasis). Further, amongst its conclusions is the following acknowledgement:

---

[1] The exceptions can be found in reports which emanate from policy advocacy organisations: such as the enthusiastic reports sponsored by the ideologically free-market American Enterprise Institute (2013), where the lead author has a policy background, as well as the related Copenhagen Consensus Center report on geoengineering (Bickel and Lane 2009), and oppositional reports associated with The ETC Group (2010; ETC Group/Biofuelwatch 2017).

The greatest challenges to the successful deployment of geoengineering may be the social, ethical, legal and political issues associated with governance, rather than scientific and technical issues. (2009: xi)

The point being made is that there are more than scientific issues involved in geoengineering. But the parallel implication is that there is 'governance' on the one hand, and 'scientific and technical issues' on the other, and that only the former has ethical/social/legal/political issues, whilst the latter, by implication, is values and politics-free. Similar formulations can be found in the IPCC expert meeting, with its focus on how various geoengineering technologies might be better evaluated; and in the IPCC 5th Assessment Reports (2013, 2014a). Indeed, the importance of integrated assessments is widely acknowledged as these are held to be the way of taking the social into account.

The NRC report concludes that "understanding of the ethical, political and environmental consequences of an albedo modification action [their preferred term for SGE] is relatively less advanced than the technical capacity to execute it" (2015a: ix). The NRC states in the summary of its report that it intends "to provide a thoughtful, clear scientific *foundation* that informs ethical, legal, and political discussions surrounding these potentially controversial topics" (2015a: 1, my emphasis). In short, the scientific is regarded as foundational and, implicitly, value-neutral. The extra-scientific is acknowledged but it is 'extra'.

Another example arises from looking at what knowledge counts when trying to understand how SGE would be likely to change the future climate. Remarkably little use is made of historical evidence of the climate effects of massive volcanic eruptions in centuries past, despite SGE's 'proof of concept' drawing on the volcanic analogue.[2] Instead, the institutional reports all rely heavily, for their assessment of the likely effects of SGE, on climate modelling work. Much of this has occurred under the auspices of the Geoengineering Model Intercomparison Project (GeoMIP) which builds on existing climate models. GeoMIP included some highly idealized modelling of a uniform reduction in the solar

---

[2] The historic evidence reveals some devastating effects, including massive crop failures, famine, war and deaths running into the millions, such as occurred in the wake of the Tambora eruption of 1815 (Wood 2014).

constant. This, in effect, mainstreamed SGE modelling, attached it to the modelling technology upon which climate policymaking is already very reliant, and suggested "that effective climate control, involving the finetuning of the amount and timing of stratospheric aerosol injection, is a potential to be realized" (Van Hemert 2017: 87).

All modelling has limitations—data gaps, over-simplification of real world dynamics, a dependence on underlying assumptions and so on—but the limitations of existing physical climate models when they are applied to SGE are particularly stark. Indeed, the NRC report acknowledges that "models provide an incomplete and imperfect picture of the world" (2015a: 25), and that there are particular weaknesses in current modelling of both clouds and aerosols (2015a: 6)—both critical for modelling SGE. It goes on to say that almost all the model results

> … are based on a limited set of idealized studies, many of which considered dimming the sun instead of actually representing atmospheric aerosols … All such simulations are greatly simplified compared to the real world, and further work is required to reduce the uncertainty in these projections. (2015a: 73)

The call for 'more research' being needed sounds unexceptional. However, research into geoengineering modelling suggests that accommodating uncertainty and making the models more complex does not appear to increase their accuracy (Hansson 2014). Relevant, too, is the distinction made by Curry and Webster (2011) when analysing climate science more generally, between epistemic and ontic uncertainties. Whilst the former are reducible, at least in theory, by more research, the latter are associated with the inherent randomness and variability of complex systems.

Putting aside the many weaknesses of SGE modelling, it is important to remember that climate models focus on modelling the physical aspects of the climate system, and with emphasis on understanding likely future temperature and precipitation variations. With SGE there is a need to understand not only the climatic impacts but also the likely social effects. Unfortunately, the tendency is to read these directly from the models—for example, more rain here will lead to more crop production. By reading social effects from simplified physical models, the "more research" which is pursued tends to be neither socially or historically informed, nor locally grounded. Essentially ignored in the major studies are things like social values, ideologies, and the adaptability or resilience or vulnerability of (complex) communities in particular situations.

This privileging of the scientific/technical is further evidenced in the ubiquitous distinction which is made in the institutional reports between Solar Radiation Management (SRM) and Carbon Dioxide Removal (CDR) geoengineering techniques. This is a far from obvious way of differentiating geoengineering technologies.[3] Relevant, for now, is that the SRM/CDR binary sets up both climate and SGE in a largely physical and mechanical manner—the climate as being like a "heating system with two knobs" (NRC 2015a: 27) where one is SRM and the other is CDR—with the implication that the expertise of scientists and engineers are primary. To this binary is added the practice whereby factors such as cost, effectiveness, and risks of deployment are regarded as central to any comparative assessment or evaluation. Here, particular forms of economic and engineering knowledge are imagined to be needed, using methods able to generate numerical values commensurable with the calculative emphasis of the physical sciences, numbers able to be integrated into climate models. As Szerszynski and Gallaraga have pointed out: "… certain disciplines are given the task of problem definition and others—typically the social sciences—are allocated the task of filling in gaps within that given frame" (2013: 2817; see also Heyward and Rayner 2013).

Three related effects flow from the knowledge hierarchy in the institutional reports, and each acts to normalise SGE and to reinforce the Imperial imaginary. Firstly, the presentation of geoengineering options emphasises their technical efficacy (will it temper global warming?) and their cost. This makes SGE appear to be the 'best' option, as indeed it probably is if these are the primary criteria. Figure 5.1, taken from the Royal Society report is the most obvious example of this move (2009: 49). This presents various geoengineering alternatives, complete with fictitious error bars suggesting scientifically credible findings and confidence about

---

[3] For example, placing the emphasis on the scale of the technology and whether it is contained or unbounded (out in the open) would result in SGE and ocean fertilisation being in one category and CCS and painting roofs white in another.

5 Knowledge-Power-Values   169

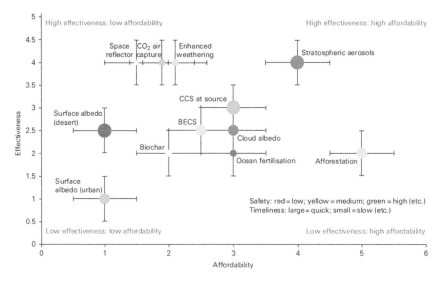

Fig. 5.1 Royal Society summary evaluation of geoengineering techniques (2009: 49)

the range of uncertainty. In the text of the report it is noted that, to avoid confusing the diagram, "the error bars are not really as large as they should be" (Royal Society 2009: 49). Indeed, the error bars are identical for each technology, which suggests they do not represent actual error ranges but merely serve to make the diagram look scientific and therefore authoritative. Only SGE (which the report labels 'stratospheric aerosols') makes it into the quadrant where techniques are considered both effective and affordable. Some 'social' considerations can be said to be included indirectly in the sub-criterion 'Safety': the report notes that SGE is "… coloured amber, because of uncertainties over its side-effects" (2009: 49). The NRC report likewise does a largely technical assessment comparing the stratospheric aerosol technique (SGE) with the technique of marine cloud brightening. A judgement is made giving one of three possible answers to a series of questions, and with each judgement allocated a 'high', 'medium' or 'low' level of confidence. Here, too, the questions focus on efficacy (stratospheric aerosol's ability to cool), availability (its technological readiness, and the time needed to develop and deploy it) and cost. Environmental and socio-political consequences are touched upon, but this is limited to noting (at a 'high' level of confidence) simply that there would be global and multinational consequences (NRC 2015a: 116–20), and it is left unstated what these consequences are and whether they would be welcome or unwelcome.

Secondly, the knowledge hierarchy shapes the way in which expertise from other disciplines is engaged. Where law is engaged it is as legal expertise and is typically given the task of assessing the extent to which existing regulatory regimes apply to geoengineering, and what existing regimes it might be best attached to (for example Armeni and Redgwell 2015). Where economics is brought in, only the most narrowly technical, cost-benefit focussed economic traditions are relied upon, which are those most amenable to quantification and being added into integrated assessment models.[4] Where political and 'justice' questions arise then power is typically treated as an external consideration rather than as intrinsic to the idea of geoengineering itself. It is operationalised as a need to model the predicted impacts on the world's poorest and most vulnerable, or as requiring attitudinal studies regarding how geoengineering or any "deploying nation or group of people" is or might be perceived (for example NRC 2015a: 135–6).

Thirdly, when climate modelling becomes the primary way of 'knowing' the future, and a highly unreliable one in the case of SGE, it crowds out "other ways of envisioning the future" (Hulme 2011: 266). As Hulme argues, "model-based descriptions of putative future climates" gain "disproportionate discursive power" and this "transfer of predictive authority", which occurs almost accidentally, can be thought of as "epistemological slippage" (2011: 256). "The future is reduced to climate", even as it becomes increasingly evident that competing and highly-contested visions and imaginations of the future will shape the impact of climate change at least as much as physical changes in the climate (Hulme 2011: 264; see also Castree et al. 2009; Castree 2015).

The idea of geoengineering is a particular outgrowth of the climate sciences and so perhaps the secondary role it allocates to other fields should not surprise. But the effects are to prioritise particular understandings of climate and climate responses which privilege technical, rather than sociotechnical, ways of addressing the climate challenge.

---

[4] Indeed costing is often limited to estimating the direct expenses associated with delivering aerosols into the stratosphere (McClellan et al. 2012).

Managing Uncertainty

The NRC report into SGE notes that it is "…impossible today to provide reliable, quantitative statements about relative risks…" (2015a: 7). Throughout, it appears torn between positing further research and a risk governance system as the way forward, and acknowledging that any decisions to deploy SGE "… will ultimately involve values and relative acceptability of various kinds of risks, factors that are outside the scope of science" (p. 146). The Royal Society report notes that a precautionary approach should apply where "the impacts of geoengineering on the environment are not yet fully known but believed potentially to be serious, if not irreversible" (2009: 38).[5] It acknowledges that geoengineering technologies are so embryonic and knowledge so limited that this may be "… a situation of 'indeterminacy' (or 'ignorance') rather than risk". But it appears uncomfortable with indeterminacy when it expresses the hope that "[t]hrough research, and the accumulation of empirical evidence, uncertainties can sometimes be recast as risks and expressed as probabilities" (2009: 37–8).

The pattern in the institutional reports is to acknowledge ignorance and uncertainty in relation to SGE, but then to understand this in limited terms or as something solvable. This manifests as reverting to, or recasting, uncertainty as, risk. A text search of the institutional literature which constituted my archive revealed the word 'risk' occurring thousands of times! The term 'uncertainty' (and its variants) occurred far less frequently and was generally attached to acknowledgement of an inability to quantify risk adequately. The implications, epistemologically, of acknowledging radical uncertainty are avoided. So too is the possibility that the complex non-linear climate system may never be sufficiently 'knowable' to make even half-way reliable predictions about the effects of deploying SGE.

---

[5] The precautionary principle is itself a controversial one and has been formulated in a variety of ways. The version most commonly used in a policy context is that formulated by the United Nations in 1992: "where there are threats of serious or irreversible damage, lack of full scientific certainty shall not be used as a reason for postponing cost-effective measures to prevent environmental degradation" (United Nations 1992). Here the emphasis is on enabling action. The principle is often also invoked to prevent or delay the utilisation of technologies or substances whose effects are believed to be harmful although not sufficiently understood (e.g. GMOs). See Whiteside (2006) for a general account of the principle.

In the institutional assessments the language of 'risk' functions to normalise SGE as a manageable object. It does this by treating risk in the narrow sense of actuarial likelihood, and by limiting the meaning of 'ignorance' to that which is knowable, but simply not yet known. If the risk can be calculated then it can be placed on the scales of costs and benefits, each with known probabilities. And it can be managed. But what if both the climate system and the effects of deployed SGE can *never* be sufficiently known to make a meaningful calculation of risk? What if it is indeterminable?

Wynne, working from a Science and Technology Studies (STS) perspective, distinguishes 'risk' (where the odds are known and quantifiable), from 'uncertainty' (with known system parameters but unknown odds), from 'ignorance' (where we don't know what we don't know), from 'indeterminacy' (which intersects with the first three but captures contextual socio-political factors as well as the conditionality of knowledge) (1992). Stirling (2010) draws on Wynne when he writes about risk, uncertainty and ignorance, but rather than 'indeterminacy' he uses a fourth category of 'ambiguity': "when experts disagree over the framing of possible options, contexts, outcomes, benefits or harms" (2010: 1030). Rayner's insightful paper, 'To Know or Not to Know?' discusses the rhetorical deployment of ignorance in relation to SGE. He notes the tension between SGE's proponents who argue that 'conducting some research is the only way to reduce ignorance about the technology', and opponents who rely on a two-fold argument: "Ignorance is a binding constraint—'we simply cannot know'—and ignorance is a source of virtue—'it saves us from folly'" (Rayner 2014: 12).

By presenting 'ignorance' and 'indeterminacy' as variants of 'uncertainty', and thereby reducible through further research to measurable risk, the institutional literature places itself in a conundrum. Especially where some variant of the precautionary principle has been embraced, as it is in the European-based studies, this effectively blocks SGE from proceeding. If SGE's deployed effects are indeed marked by radical uncertainty and un-knowability, then the risks must remain unknown and potentially dangerous, and the precautionary principle acts as a presumption against proceeding.

This line of argument elicits at least two responses. The first is to invoke precaution in relation to both the climate change problem anticipated *and* the proposed technological solution of SGE, but to argue that the climate risk trumps the technology risk. A second line of argument is to acknowledge the deep uncertainty and treat SGE as a practical challenge to be resolved through trial and error. David Keith's approach comes close to this (2013). He makes the case for commencing by injecting stratospheric sulphate aerosols in minimal quantities, then slowly ramping the dosage up (or down), all the while monitoring the climate system and the actual effects and feedback loops observed. But no institutional report currently endorses this approach.

It is an approach that has some commonalities with concepts such as "post-normal science": where "facts are uncertain, values in dispute, stakes high and decisions urgent" (Funtowicz and Ravetz 1993: 739). There is commonality, too, with its implications: that since uncertainty cannot be eliminated, it must be managed, and that values must be made explicit rather than be left assumed. Whilst a number of academic papers situate geoengineering proposals within the realm of post-normal science (for example Bellamy et al. 2012: 610), the concept is not embraced in the institutional literature or in the IPCC more generally (see Krauss et al. 2012). To do so risks challenging the prevailing assumptions and claims about how truth claims are made and knowledge and 'consensus' arrived at.

Counting Without Taking Account

In the knowledge hierarchy that prevails in the institutional accounts and assessments of SGE, economics occupies the rung below science and engineering. Most of the reports devote some space to analysing what Scott Barrett, who led the economics discussion at the IPCC Expert meeting in 2011, has elsewhere called the "incredible economics of geoengineering" (2008). Almost all the institutional literature includes cost estimates for deploying SGE and, as we have seen in the Royal Society example above, cost criteria are typically included as part of the assessment process.

SGE is, on the face of it, incredibly cheap, easily affordable by a single middle-income country, and even by a super-wealthy individual. The NOVIM study regards the costs as so low as to not be an obstacle (Blackstock et al. 2009). A study for the German government concluded that operational costs per watt per square metre (W/m^2) of incoming solar radiation compensated for, could go as low as $2bn per annum (Rickels et al. 2011: 57).[6] The NRC report cites McClellan et al. (2012), SGE proponents, who estimate that delivering 5Mtonnes of sulphur per annum into the stratosphere (20–30 kms up) would cost between $2bn–$8bn per annum. The report does caution that these estimates exclude a number of things: the costs of producing the aerosols, potential damage to aircraft of operating in a high sulphate environment, the cost of observing systems (such as satellites) to monitor the climate effects, and the fact that large-scale engineering projects often experience cost escalation (NRC 2015a: 78–9). But even taking these into account, it is clear that the report, too, believes the direct costs of deploying SGE are low. In its summary report the NRC concludes that SGE is "inexpensive to deploy (relative to cost of emissions reduction)" (2015b: 3), by "at least an order of magnitude" (2015b: 6).

Importantly, all the institutional reports also conclude that SGE would be much cheaper than both mitigation and unmitigated climate change. The Royal Society report concludes that the direct financial cost of deploying SGE "… is small compared to the costs of the impacts of foreseeable climate change, or of the emissions reductions otherwise needed to avoid them" (2009: 49). It cites the Stern report which estimated the costs of conventional mitigation to be in the order of $1 trillion per year (Royal Society 2009: 44). One of the main take-home messages received by policymakers and also conveyed in the media, is that SGE is cheap, and cheaper than mitigation (NRC 2015a: 3; Curvelo 2012: 190).

Calculating the direct costs of SGE is relatively easy. Whether the indirect costs are effectively captured is invariably more contentious (see also MacKerron 2014). Stilgoe argues that SGE is only cheap "… if we choose

---

[6] As a broad indicator, to balance the warming effects of a doubling of CO_2 concentrations would require a compensation of about 4 W/m^2 (Royal Society 2009: 23). To deploy SGE *now* as a pre-emptive strategy would need a smaller W/m^2 compensation in the order of 1 W/m^2.

to do some sums and not others; if we internalise some costs and externalise others ..." (2015: 87). This echoes the IPCC Expert Report which stresses that direct costs and benefits associated with implementing and operating geoengineering need to be considered alongside "cost valuations for potential social and environmental externalities" (2012: 4): an enhanced drought in one area, an increased likelihood of flooding in another, reduced crop damage in yet another. This point is also made in the RAND Corporation report. An SGE system, it notes, citing work by Goes et al. (2011):

> will appear cost-effective if decisionmakers are very certain—on the order of 90-percent confident—that it can be maintained for decades and that its adverse impacts will prove smaller than about 0.5 percent of gross world product. Otherwise, the study finds that the system's risks outweigh its benefits. (Lempert and Prosnitz 2011: 7–8)

These are weighty conclusions which, in effect, rule out SGE producing benefits exceeding the costs. However, given the high levels of uncertainty regarding the likely 'adverse impacts' from deploying SGE it is unclear how even these calculations and probabilities can be derived with any credibility. Leaving aside the NRC report conclusion that maintenance would be needed for centuries rather than decades, is it plausible to imagine decision-makers reaching '90 percent confidence' that a deployed system of SGE could 'be maintained for decades'? What does 90 percent confidence even mean in this context? Further, the conclusion rests on comparing one scarcely-known (although presumed in theory to be knowable) and uncertain future (SGE-altered) with a perhaps more predictable, but still speculative, climate changed future. This is a recipe for self-serving modelling!

Not surprisingly, other studies critique Goes et al. (2011) and come up with entirely different conclusions (for example, Bickel and Agrawal 2013). In its report, the American Enterprise Institute (AEI), an organisation strongly committed to free market capitalism and typically associated with climate change denialism, draws on these alternative studies. The AEI report argues that SGE would create net benefits of up to $10trillion in comparison to unmitigated climate change (Lane and

Bickel 2013: 13). The papers prepared for the Copenhagen Consensus report, by the same authors as the AEI report, estimate a direct benefit-cost ratio of 25 to 1 and use this to argue that it makes sense to use SGE to enable mitigation efforts to be decelerated (Bickel and Lane 2009, 2012).

Whether used by proponents or opponents of SGE, the language of "cost-benefit" is ubiquitous. The NASA workshop in 2006, for example, gives an indication of the utilitarian lens commonly used:

> In principle and under favorable circumstances, this strategy [using SGE pre-emptively] could be consistent with an economically efficient climate policy. Economic efficiency requires minimizing the present value of the sum of the damages from climate change and the costs of reducing those damages. (Lane et al. 2007: 11)

The internal logic of this position is hard to fault provided one accepts the normative centrality of 'economic efficiency' and the associated methods of cost-benefit calculations, including the belief that all costs and benefits of SGE can be converted to a single monetary measure.[7] These are heroic assumptions. At the same time, they are core beliefs of the dominant economic order and the larger climate policy debate is often framed in terms of cost. It is, for example, regularly argued that money spent on cutting greenhouse gas emissions will be repaid in reduced risk of future climate damage, and that any reduction in GDP from action would be less than the eventual reduction of GDP from inaction (see for example the IPCC 2007 and Stern 2007). If one adopts this logic then it makes sense to embrace SGE if it promises to provide an even better Return on Investment.

The language of 'cost-benefit' permeates the institutional literature, as does a methodological reliance on calculability, which in turn requires disparate things to be made commensurable by converting them into dollar equivalences. Other terms drawn from the lexicon of neo-classical

---

[7] There are many existing critiques of cost-benefit analysis, especially in relation to environmental damage (see for example Wegner and Pascual 2011). See also Porter's account of the rise of cost-benefit analysis in the United States (1995) and MacKenzie's analysis, in the context of carbon markets, of 'making things the same' (2009).

economics and the standpoint of maximising utility are also used regularly. These include framing geoengineering as a 'global public good' (or bad) as parts of the IPCC Fifth Assessment Report suggest (2014a: 1771; see also the so-called 'Oxford Principles' in Rayner et al. 2013), and understanding its potential deployment as a 'collective action problem' (see for example IPCC 2012: 24). These terms have drawn the attention of a number of commentators and are not discussed here.[8] It is sufficient to note that where economic issues are discussed in the institutional literature they are typically narrowly focussed and derived only from the hegemonic school of economic thinking in the West. They are 'economics' rather than 'political economy'. They are from the tradition of *counting* the expected dollar costs and benefits, rather than *accounting* (in the sense of holding to account).

The institutional assessments of SGE simultaneously rely on the methods, epistemologies and implicit values of mainstream neo-classical economics, and are often uncomfortable with the implication that this promotes SGE.[9] This disjuncture may help explain 'Recommendation 1' in almost all the major institutional accounts: that mitigation should remain the primary instrument for tackling climate change, and that so-called 'moral hazard' should be avoided and nothing done to remove the focus from emissions reductions. In part this is a recognition that SGE only masks the build-up of GHGs and that GHG concentrations will still need tackling at some point. But perhaps repeated concerns about 'moral hazard' may function as an indirect way of suggesting that mitigation should be prioritised despite the costs. It is an acknowledgement that the embrace of counting has resulted in an undermining, when considering climate change, of holding those causing it to account.

---

[8] See, for example, the debate on whether SGE is a 'public good' between Gardiner, who argues that doing so "arbitrarily marginalises ethical concerns" (2013: 513 and also 2014), and Morrow, who argues that "[f]raming geoengineering as a public good is useful because it allows commentators to draw on the existing economic, philosophical, and social scientific literature on the governance of public goods" (2014: 95). This suggests they are in agreement that the concept is doing normative work, whilst disagreeing about the merits of that normative work and the framings being relied upon.

[9] Only the most free-market enthusiastic policy reports, such as those of the AEI and the Copenhagen Consensus Center already mentioned, are enthusiastic about drawing out this implication.

In short, in the disciplinary hierarchy of 'official' assessments of SGE, economics plays an important role: after climate science, but well ahead of the social sciences. The focus is on the calculation of costs and the application of cost-benefit techniques. The not implausible message is that SGE is cheap, and significantly cheaper than mitigation. Notably absent is any reflection on SGE's manifest compatibility with the continuation of emissions-intensive and growth-focussed business-as-usual, a point I will return to further on. And the paradox is that whilst SGE's cheapness is highlighted, it is also generally regarded with discomfort and diverted into discussions about 'moral hazard', a point to which I will also return.

* * *

Epistemologically, the approaches adopted—the disciplinary hierarchy, the assumptions of knowledge commensurability, the possibility of knowing and modelling all that is needed, the privileging of utilitarian conceptions of benefit—are approaches widely enacted in mainstream climate policy-making. But whilst these approaches serve to normalise SGE, add weight to calls for its embrace, and make it possible for SGE to be regarded as a normal part of climate policymaking, they also, under scrutiny, make it harder for SGE to be normalised.

Szerszynski and Galarraga have argued that geoengineering research is characterised by "organised epistemological irresponsibility" (2013: 2818). This becomes apparent especially when geoengineering options are appraised. Bellamy et al. (2013) have explored a different approach to geoengineering appraisal than that adopted in the institutional assessments. They note that by opening up appraisal to a wider range of "framings, knowledges and future pathways", different outcomes emerge. Importantly, they conclude that SGE, which typically outperforms other geoengineering options, performed poorly (2013: 935).

The "subordination-service" mode, as Szerszynski and Galarraga term it, drawing on the vocabulary of Barry et al. (2008), in which disciplines are brought together, maintains a particular framing of geoengineering as a techno-scientific response, admittedly controversial, to a global warming problem.[10] It assumes that different ways of knowing can be made commensurable, and even integrated into models as data.[11] Szerszynski and Galarraga (2013) have called for new forms of interdisciplinarity—ontological, agonistic, asymmetrical—which deepen and open up to scrutiny the conversation about geoengineering. Others have spoken of "minority interdisciplinarity" and "radical inter-disciplinarity" (cited in Castree 2015).

The paradox, therefore, is that even as the centrality of values and visions is widely acknowledged to be central to thinking about geoengineering, the framing of the debate within a particular disciplinary hierarchy largely shuts this down, and proceeds to model options, costs, and effectiveness (in reducing temperature) in standard climate reductionist terms. As Szerszynski and Galarraga observe, there is little recognition that "… the fashioning of climate [is] … a form of worldmaking: a moment of historical rupture in which intentions and ways of thinking may be transformed" (2013: 2822–3).

Power

Since it aims to engineer and re-shape the global climate, SGE is, in ambition, a powerful technology. In this section I explore three manifestations of the not always straightforward relationship between SGE and the dominant political-economic order: "imperial market globalism" in Steger's formulation (Steger 2009). Firstly, I examine the implications of the invocation of urgency and climate emergency. This is an instance of discursive power, arguably with coercive effects. Secondly, I touch on the imagined relationship and likely effects of deployed SGE on the existing geo-political order. Thirdly, I explore the complicated compatibility

---

[10] It could even be argued that that the relevant knowledges for understanding geoengineering are primarily those which think about power, ideology and the co-production of the techno-social; secondarily, those on the terrain of climate policy considering the implications (and public perspectives) of adding intervention as a new policy leg; and then the relevant fields of physics, chemistry, engineering, various Earth system sciences and cognate disciplines. This is the reverse of the current order of knowledge-privileging.

[11] See also Castree's account of 'convergence science', and other attempts from the sciences to "… bring people and nature into a single analytical domain, aspiring to mirror in a computational environment real-world couplings between socio-economic and physical systems" (2015: 4).

between SGE and the capitalist economic order. Taken together, it is clear that material, coercive, discursive, institutional and structural power are entwined. They also suggest that not only is SGE a powerful technology, it is also a technology of the powerful.

Emergency and Risk

Emergency discourse permeates discussion of climate change. "'I his is an emergency", said UN Secretary-General Ban Ki-Moon, "and for emergency situations we need emergency action" (cited in Lagorio 2007). Such arguments are especially evident in discussions of SGE. Indeed, as Nerlich and Jaspal report, the emergency framing is the preponderant underlying narrative in both scientific and popular discussions of geoengineering (2012; see also Bellamy et al. 2012). Both the institutional reports and the policy-influential literature on SGE regularly refers to 'urgency', 'tipping points', 'dangerous', 'abrupt', 'rapid' and 'non-linear' climate change. Emergency framings are also prevalent in popular and press accounts of SGE. In short, the idea of "emergency" is invoked even when the term itself is not used (Markusson et al. 2014: 282; see also Horton 2015; Sillmann et al. 2015).

However, the reliance on 'emergency' takes two slightly different forms. There are those who imagine, quite literally, an event or set of events occurring which would require the deployment of SGE. And then there are those who argue for SGE to be developed and possibly deployed in order to reduce climate risk, because a climate emergency in a *generalised* sense exists.

The literal emergency rationale typically imagines the technology being readied for deployment in the event of a climate crisis, as a fast-acting antidote (ideally temporary) to tackle, say, 'runaway climate change'. The Royal Society report, for example, argues that SGE "… may provide a potentially useful short-term backup to mitigation in case rapid reductions in global temperature are needed" (2009: 59).

Whilst emergency discourse is often mobilised, what would constitute an emergency is rarely made clear. The NOVIM report, *Climate Engineering Responses to Climate Emergencies*, imagines as an emergency the situation where "the impacts of a warmer world on humans and natural ecosystems ... [are] more severe than current median predictions" (Blackstock et al. 2009: 1): a strange combination of the language of "median" and "emergency". In a report whose overwhelming focus is on climate emergencies and SGE, it defines climate emergencies as "those circumstances where severe consequences of climate change occur too rapidly to be significantly averted by even immediate mitigation efforts" (2009: 1). The solution to such vagueness, according to the GAO report, was more data on climate thresholds and "a way to determine when a 'climate emergency' is reached" (2010: 16). For the chair of the US House Science and Technology Committee this is a decision for experts: "the global climate science and policy communities should work towards a consensus on what constitutes a 'climate emergency' warranting deployment of SRM [SGE] technologies" (USHCST 2010: 40).

On the few occasions in the institutional literature where a particular emergency is envisaged it is typically imagined as an event. The IPCC Expert report notes that the:

> most discussed of the possible emergencies have been a methane burst as a result of the rapid thawing of permafrost and/or clathrates trapped in the sediments of the continental shelves, the rapid loss of ice mass from the Greenland and/or Antarctic ice sheets, collapse of the Amazon rainforest, or greatly accelerated, runaway warming. (2012: 55)

But what, for example, would indicate the collapse of the Amazon rainforest, or indeed signal its impending collapse, and be clearly not attributable to, say, logging or agricultural conversion? And how much more rapid than it currently is would Greenland ice loss need to be? To pose such questions is to suggest their unanswerability.

The NRC report, which generally eschews the literal emergency rationale, mentions one scenario which they deem 'hypothetical but plausible':

> If, for example, global warming resulted in massive crop failures throughout the tropics (e.g., Battisti and Naylor 2009), there could be intense pressure to temporarily reduce temperatures to provide additional time for adaptation. (2015a: 32)

Here one can detect echoes of the Imperial imaginary of SGE, especially its Salvation strand. But who would exert this pressure and how? And who would decide it warranted deployment of SGE? And from a purely evidentiary perspective would, for example, the massive crop failures and droughts of 2015/16 count? And how would the global warming effect and the El Niño effect be disentangled?

To date the only specific call for an event-linked use of SGE has come from the Arctic Methane Emergency Group (AMEG), a grouping of Arctic scientists and others. They argue that the emergency is already upon us and have called for immediate geoengineering to re-freeze the Arctic (AMEG 2014). But they have found few supporters for their call, even though the emergency rationale is embraced by many, and the evidence for the Arctic having crossed a tipping point of some sort is compelling (Lenton 2012).

In short, the emergency argument lacks precision even whilst it appears to envisage a threshold determined by scientists being crossed, leading to the deployment of SGE. Perhaps the clearest implicit threshold can be attached to the 2°C target contained in the Paris climate agreement with its suggestion that this is the zone of "dangerous climate change" (UNFCCC 2015). Whilst the Paris agreement makes no mention of SGE, there are increasing calls from SGE's enthusiasts to incorporate it: "the fact that the Agreement contains explicit quantitative temperature goals instead of, say, an explicit carbon budget goal ... actually facilitates the eventual inclusion of [SGE] in the post-Paris system" (Horton et al. 2016: 5).

A further difficulty with the emergency argument can be derived from the work of a number of leading climate scientists. Tim Lenton has cast doubt on whether a precise 'tipping point' in any major physical component of the climate system could be predicted or observed in time to be reversed: "by the time you detect that abrupt change is either imminent or underway, the tipping point may long since have been passed and the change simply cannot be reversed" (2013: 1; see also Sillmann et al. 2015). Any element of the climate system which has 'tipped' is on its way to an alternative state. Whilst in principle reversible, "it could demand a reduction in radiative forcing well below the pre-industrial level" (Lenton 2013: 2): in practice, this would entail an especially high dosage of sulphate aerosols into the stratosphere to achieve the desired effect. Since the concepts of 'non-linearity' and associated 'tipping points' is the most scientifically plausible justification for an 'emergency', the fact that these could only be detected after the event and then be irreversible, would appear to be a fatal flaw in the argument. Lenton does acknowledge that SGE may be viable "… for a subset of tipping points", fast-responding systems "that are directly related to temperature change": notably Arctic sea-ice (not land-ice) or monsoons. In the case of monsoons he notes that they "are particularly sensitive to aerosol forcing and may actually be disrupted rather than protected by deliberate aerosol injections" (2013: 3). Modelling work by McCusker et al. (2012) casts doubt on the ability of SGE to prevent polar climate emergencies. In short, of all the emergency triggers envisaged earlier, none remain scientifically credible as 'solvable' by SGE. In its literal version, the emergency notion appears to be internally incoherent and impossible to operationalise, a reality now widely recognised, even by some sympathetic to the development of SGE (Horton 2015).

Increasingly, proponents of what I call the Imperial imaginary of SGE rely on both a generalised sense of a climate emergency to legitimate SGE's deployment, and on a climate risk reduction rationale to argue for pre-emptive deployment. Michael McCracken, in his participant presentation to the IPCC expert meeting on geoengineering, notes that an implicit assumption of the emergency formulation "… is that climate is reversible, and this is not at all clear. In addition, adaptation is likely to have spread out the range of optimal temperatures for various societal and environmental systems, such that a sudden, sharp cooling might be very disruptive" (IPCC 2012: 55). For McCracken this is reason to deploy SGE as a precautionary measure, a prudent action, prior to any 'tipping'. Keith argues for research leading to the "wise use of geoengineering" (2013: 174) as a preventative technology aimed at flattening the peaks of global warming and slowing the rate of increase of our current temperature trajectory. His essential argument is that in the context of a generalised climate emergency, SGE should be used pre-emptively at a moderate dosage. For both McCracken and Keith, SGE will take us to a new and safer climate, not restore a previous climate or aim to maintain an optimum global temperature. This approach avoids the manifest incoherence in the literal emergency argument, whilst retaining the general framing of a climate emergency.[12]

These are clearly more plausible propositions than those contained in the literal emergency version. But the risk framing still requires a generalised sense that we face a climate emergency in order to justify an intervention such as SGE, and the coercive work that risk does in this situation cannot be ignored. As Ulrich Beck, the pre-eminent contemporary theorist of risk, has argued:

> Risk society means that the past is losing its power of determination of the present [and] being replaced by the future … as the basis for present day action. We are talking and arguing about something that is not the case, but could happen if we do not turn the rudder immediately. *Expected risks are the whip to keep the present in line.* The more threatening the shadows that fall on the present because a terrible future is impending, the more believed are the headlines provoked by the dramatization of risk today. (1997: 20 my emphasis)

'Emergency' is fundamentally a social category, although it is not treated as such in the institutional reports on SGE. Who declares this emergency and thereby triggers the deployment of SGE, and at what spatial scale? And for whom is it an emergency? As Hulme has observed, from a social perspective, emergencies never simply happen, they are always declared (2014: 21–6). Calhoun, in his work on the emergency imaginary, has described 'emergency' as "… a way of grasping problematic events, a way of imagining them that emphasizes their apparent unpredictability, abnormality and brevity, and that carries the corollary that response—intervention—is necessary" (Calhoun 2010: 25). Since it seems clear that SGE cannot prevent or reverse climate emergencies, then the effect of relying on emergency rhetoric can only be to mobilise a response in favour of SGE, to act as a call to action. Further, by suggesting we are in exceptional times, it is implied that exceptional interventions are needed, perhaps even that a 'state of exception', in the Schmittian

---

[12] This argument has its own assumptions. It requires climate risk to be prioritised over any risks which may flow from using SGE technology. It assumes that an engineered reduction in global average temperature rises will, on balance, be positive, and certainly better than continued global warming.

sense, exists. This is coercive rhetoric and an assertion of power. Perhaps Clive Hamilton is correct to suggest that SGE might be the dictator's technology of choice (2013). Or it may be understood as a technology of our securitized times.

The temporal dimension of SGE is intriguing in other respects too. Unlike in Calhoun's formulation there can be no suggestion that the climate emergency will be brief. On the one hand SGE is imagined as a temporary technology, to deal with an emergency. On the other hand it is widely acknowledged that embarking on SGE is likely to be a centuries-long commitment or at least a multi-decadal one (see NRC 2015a). Troublingly, therefore, the suggestion is that both the climate emergency and the SGE intervention are unlikely to be brief, and that we may be entering a permanent emergency, if that is not a contradiction in terms, a 'state of exception' as the normal state of affairs. Risk, as Beck has argued, is not only chance and side effect. It can also be understood as "the way of being and ruling in the world of modernity" (Beck 2006: 330). In SGE, risk and emergency come together to produce power effects.

In short, the literal emergency argument for SGE is incoherent and unpersuasive. But 'climate emergency' as a generalised framing does legitimating work by making SGE appear both necessary and vital: who could oppose 'action' when faced with an emergency? Both 'emergency' and 'risk' are coercive rhetoric, and assertions of power. This may help explain why there are no calls for SGE coming from the global South, even as the climate predicament of the global South is regularly cited as grounds for deploying SGE.

Global Order

By any measure, the deployment of SGE would have global climatic consequences and effects which cross existing national boundaries. Consequently, any consideration of SGE must inevitably consider the international and geo-political implications, and it must do so in a context in which the US is the dominant global superpower, one prepared to use its muscle to assert its interests. This remains true even as challenges to its economic dominance have emerged. Two tropes emerge repeatedly in the institutional assessments of SGE. Firstly, the fear of unilateral deployment and the related question of 'who decides' on the use of

SGE. Secondly, and as a solution to this, an emphasis on getting suitable governance structures in place. Both operate at the intersection of climate policy and geo-politics. In the institutional reports they are typically handled from the standpoint of the already powerful.

The Bipartisan Policy Center report calls for the US to play 'a pivotal role' in engaging with 'other major countries' on SGE policy (BPC 2011: 28), although which countries these are is left unspecified. A paper on 'Unilateral Geoengineering' was at the centre of a workshop convened by the influential US Council on Foreign Relations in May 2008, and attended by a range of SGE policy-influential individuals. This paper stressed the need to build norms for 'responsible geoengineering' much like those governing first use of nuclear weapons or the safe testing and deployment of GM crops, and to roll these norms out globally. But it expressed anxiety about norms constraining 'responsible' countries, implicitly the US:

> Aggressive norms might stymie research on geoengineering within countries that are most likely to honor the norms—that is, the countries that are most likely to give closest attention to possible collateral damage from geoengineering—while doing little to thwart dangerous unilateral geoengineering by countries and institutions that care less about international norms. (Ricke et al. 2008: 12)[13]

The NRC report also expressed concerns about "unilateral" deployment of SGE (2015a: 122ff). Although not made entirely explicit, in the NRC report 'unilateral' appears to mean any deployment not conducted by or coordinated with the United States. It should be remembered that the US security establishment played a leading role in the NRC report as both a funder and a participant. In an echo of the post-2001 discourse around 'weapons of mass destruction', the terms 'irresponsible', 'uncoordinated' and 'rogue' deployment are repeatedly used, as well as the need for ongoing satellite capability able to both detect unilateral deployment and monitor SGE's climate effects if deployed non-unilaterally. It suggests supplementing this surveillance capability with intelligence gathering

---

[13] Few globally, especially in relation to climate policy, would regard the US as a 'responsible' country, even more so in the Trump era.

"on the movement or use of albedo modification agents (e.g., chemical feedstock transport, manufacturing, injection facilities)" (NRC 2015a: 111–13).

A July 2016 speech by John Brennan, the head of the US Central Intelligence Agency (CIA) suggests he is paying close personal attention to SGE, seeing it as a "beneficial advance" but one which might have "destabilizing effects in the long run". He appears to regard SGE in a positive light as a means to limit temperature increases and some of the associated risks, and give time to transition from fossil fuels. SGE's cheapness is also attractive, and here he cites the NRC report's cost estimate of $10bn per annum, but fails to cite the many caveats which surrounded this estimate. For Brennan the downsides of SGE include that greenhouse gas reductions would still be required to address ocean acidification, that it "could trigger sharp opposition by some nations", and an absence of "global norms and standards … to guide the deployment and implementation" (Brennan 2016: n.p. for all the above). Brennan's perspective is not unique within the US security establishment. A 2016 study by Engelke and Chiu, for example, entitled *Climate Change and US National Security*, repeatedly refers to geoengineering in the context of a broader plea to "make the climate security concept part of a comprehensive narrative tying climate insecurity to the United States' core national interests" (2016: 2). In the more general securitisation of environmental and climate policy geoengineering is playing an increasingly prominent role.

Concerns have been raised that a low-lying island (say, the Maldives) or a super-rich individual might unilaterally embark on SGE. But it is hard to imagine unilateral deployment of this sort without the prior approval of the US. Nevertheless, action by other global powers is conceivable. Healey and Rayner provide a plausible example of the potential for deployed SGE to generate inter-state conflict. They ask us to imagine India having conducted SGE experiments shortly before the devastating 2010 floods in Pakistan. "It is hard to imagine that members of the Pakistani public, and certainly Pakistani politicians would not have held India responsible for the damage that had been caused by the floods… [especially given that] attribution for climatic events is unclear" (Healey and Rayner 2015: 14). Both countries, of course, have nuclear weapons.

What I am arguing is that in the institutional assessments of SGE and the policy-influential literature, when unilateral deployment is imagined as a problem, the concerns emerge from the standpoint of the already powerful, from those who imagine deploying it themselves. Deployment by the US is not considered to be 'unilateral'. Discussion around this is sometimes presented as who decides when, where and how to deploy SGE?[14] The Royal Society report noted that, "[a]lthough the analogy is flawed, some commentators have asked 'Whose hand will be on the global thermostat?'" (2009: 40). The issue is commonly raised but rarely addressed in detail in the institutional assessments.

In general, the European reports tend to imply that some form of global, international governance is needed, whilst the US-based reports do not. A number of writers, generally European, take a nominally universalist view and argue for some form of global governance/regulation. For legitimacy this should involve, at least in theory, all the nations of the world (see for example Virgoe 2009). The Royal Society report suggests that perhaps the UN or one of its agencies should govern geoengineering (2009: 60). Others have proposed the development of a geoengineering protocol to be attached to the United Nations Framework Convention on Climate Change (Scott 2013). It is hard to see the suggestion that *all* the nations of the world might voluntarily agree on a planetary thermostat and SGE as anything other than naïve.

US-based assessment reports generally avoid addressing the issue directly or else limit their focus to research governance rather than the governance of deployment (see for example the NRC report 2015a). It is more informative, therefore, to look at the policy-influential literature. David Victor urges the United States to oppose any attempt to regulate or govern globally on the grounds that it would in practice result in a ban on geoengineering and geoengineering research (2011). He does acknowledge however that as a matter of necessity, some form of consent might be required from a few other "countries that matter" (Victor 2011). Lane and Bickel, in their work for the American Enterprise Institute (2013), which is less constrained than their work for the institutional assessments,

---

[14] Importantly, the question of whether to deploy SGE at all gets less attention and attempts to get SGE prohibited (under the Convention on Biodiversity for example), have been strongly resisted.

explicitly reject any notion of global governance, arguing instead "... for a coalition with enough bargaining power to impose its preferences over global climate" (2013: 20). Whilst the AEI is an outlier more generally in policy debates, the 'climate policy realism' expressed here is perhaps the dominant view in the US-based policy-influential literature, although it has struggled to find formal expression in the institutional literature.

In the broader academic literature on SGE there are some, most notably UK-based geographer Mike Hulme, who argues that "a planetary thermostat in the stratosphere would be ungovernable" (2014: 86). In part this is because claims that the climate can be mastered and managed are illusory, and SGE as a technology purporting to be able to do so is unreliable. In Hulme's case this is an important reason to reject SGE and even applied research into SGE. Where the view that developing and deploying SGE should be the decision of a handful of nations is expressed it feeds the fears, especially in the global South, that SGE will mean a coalition of the powerful imposing their desired climate on the majority, and perpetuating the existing, unequal, global order.

Capitalism

The dominant economic system of capitalism is associated with modes of extraction, production and disposal which are environmentally-intensive—what has been described as the "take-make-waste" system (Doppelt 2012).[15] The rapid expansion of output (or GDP) growth, built on fossil-fuel energy, has driven the sharp rise in anthropogenic emissions of CO_2 and, consequently, global warming. Some have concluded that tackling climate change requires the assumptions and practices of capitalism itself to be confronted (for example Klein 2015). The more mainstream response argues that reforming and 'greening' capitalism by de-coupling it from environmental degradation will suffice (for example World Bank 2012).

---

[15] It is true, of course, that similar, often higher, levels of environmental degradation could be found in the planned economies of the former Communist states, and that environmental degradation is also a feature of nominally Communist, state-capitalist, economies such as contemporary China.

Mitigation strategies, aimed at reducing emissions, are the currently preferred forms of climate policy. But they require significant changes to energy and transport systems and to patterns of land-use. This takes time and requires new coalitions of support to be built and existing vested interests to be overcome, both nationally and internationally. Mitigation is also costly—one estimate is that it would cost an eye-watering $124 trillion to shift to the renewable and clean energy profile implied in the Paris agreement (Jacobson et al. 2017). To date, 'green growth' is proving elusive. Whilst there is some evidence of greater eco-efficiency and relative de-linking of the relationship between growth and emissions, there has not been the absolute de-linking (GDP growing with GHG emissions declining) which is a pre-condition for stabilising GHG concentrations at a lower level than they are currently.

Kevin Surprise, writing from an ecological Marxist perspective, has argued that SGE offers a "spatio-temporal fix" to the second contradiction of capitalism, in that it can "elongate the timescales within which green capitalism proves effective" (2018: 1240).

> … if climate change is the most expressed manifestation of the second contradiction and green capitalism the central crisis management strategy, then [SGE]… can defer the onset of this contradiction. This would allow not only for timely effectuation of green capitalism and the deferral of climate crisis but the expansion of capital accumulation in an otherwise finite system. (2018: 1240)

Turning to the institutional assessments, perhaps most notable is what these do *not* say, their silences. The studies all assume that the seriousness of the climate situation is such that a radical intervention, SGE, *may* be needed to re-shape the planet's climate. In their account, the driver of climate change is increased emissions and rising levels of atmospheric GHG concentrations. At best it is assumed that fossil fuel use is implicated. But no consideration is given to the socio-economic drivers of these physical and technical factors. The institutional reports are silent about the dominant model of growth and consumption that has created a climate situation of such seriousness. Capitalism is nowhere directly addressed. Perhaps it is alluded to in the German Environmental Agency report which notes, not with approval, that with SGE "little or no behavioural change has to be demanded from society for the reduction of CO_2 emissions" (Ginzky et al. 2011: 41). And the NRC report briefly makes a similar point, although without disapproval: "Implementing an aggressive program of emissions abatement or mitigation presents major challenges to how we live and function as a society" (NRC 2015a: 15).

In the institutional reports SGE is repeatedly acknowledged to be an environmentally masking technology—one that deals with the symptom of planetary warming, not the physical drivers. All the reports acknowledge it does nothing to reduce GHG concentrations, ocean acidification and so on. But no mention is made that it may be a masking technology in other ways: that it masks, conceals or postpones the need to shift from the current patterns of extraction, production and consumption, and the associated pursuit of endless growth. Certainly, by offering the possibility of masking the warming effects of rising GHG concentrations whilst enabling the existing economic patterns to continue, SGE can be seen as compatible with the dominant economic logic of our times. The institutional reports are silent in this regard and thereby presume the continuation of existing power relations and the dominant economic order.

The question of capitalism does come up in the policy-influential literature. As I have shown in the previous Chapter, some policy think-tanks of a libertarian bent (such as the American Enterprise Institute and Lomborg's Copenhagen Climate Consensus) and a few politicians (such as Newt Gingrich) have explicitly embraced the compatibility between SGE and capitalism. They make it explicit that they find SGE attractive because it allows "politically bankrupt", expensive and growth-reducing mitigation to be avoided (Lane and Bickel 2013: 20). Indeed, SGE is probably cheap and almost certainly cheaper than mitigation, especially if cheapness is understood as a monetary measure of cost (Smith and Wagner 2018). And the internal logic of this position is undeniable. Cost-benefit and market-friendly approaches have been central to climate policy discourse and practice since at least the Kyoto agreement, and so it is hard to see why policymakers working within such a framework should not choose the cheaper option, especially if its benefits outweigh its costs.

However, other policy-influential writings, also from within the Imperial imaginary, are less attracted to this logic. If SGE is seen as an intervention to slow the rate of warming and extend the time for a green transition, then cheapness is an added attraction rather than a motivation for action. Keith, for example, argues:

> I embrace this [utilitarian, cost-benefit] framework for making near-term policy trade-offs about air pollution or tax policy, but I think this analytical machinery is far less useful when pressed into service to make global-scale all-but-irreversible decisions that span centuries. (2013: 165)

However, elsewhere he argues that cost considerations are significant in relation to climate adaptation. "By reducing the needed rate for adaptation, [SGE] could reduce economic costs of adaptation" (MacMartin et al. 2014: 2)

Keith also addresses the question of capitalism directly when he argues that:

> ... environmentalists have conflated two different causes: repealing the excesses of aggressive capitalism and minimizing harm to the environment ... [which] while they are connected, I don't see a tight one-to-one linkage between them. (2013: 143)

Keith accepts there have been difficulties in minimising harm to the environment. But this is "... not because political and economic liberalism are inherently anti-environmental, but because an accumulation of [corporate] power and private political money has frustrated the ability of government to act in the public interest" (2013: 148). In this view, rather than attributing a role for liberalism in that accumulation of money and power, it is understood that the ideal of independent and public-spirited government may have been corrupted. This leads him to ask, rhetorically, "must we fix capitalism in order to fix the climate?" (2013: 143). 'No' is the implied answer, although Keith acknowledges the need to address some of capitalism's 'excesses'. It is the mirror response to Naomi Klein's argument in *This Changes Everything* (2014).

Elsewhere Keith asks, again rhetorically, "should we prefer social fixes to technical fixes?" (2013: 147). The suggestion is that what the climate problem lacks is a mechanism or mechanisms to 'fix' it. The question also implies that the social and the technical are distinct although numerous studies reveal that they are always entwined (see for example Pinch and Bijker 1987), indeed that in the emergence of new technologies the social and the technical are imagined together, and that technical 'fixes' are socially shaped and have social effects.

When it comes to SGE and capitalism we are left with something of a paradox. Whilst individuals amongst the political elites and the super-rich have shown interest there is little evidence of broader enthusiasm for SGE among climate policymakers, in UNFCCC talks and similar intergovernmental encounters, or in elite forums such as the World Economic Forum. Given the compatibility between the maintenance of the dominant order and the technology of SGE, and given the cost of existing mitigation strategies, and the potential for climate change to have damaging social and economic effects, why has governmental and institutional enthusiasm for SGE failed to materialise? I will return to this issue in the concluding Chapter but for now I raise a few tentative and speculative points to help in considering this question.

Firstly, there are few vested interests. There are some, but not many, business interests (such as patent holders or companies hoping to benefit from new technologies) but no business lobby for SGE. It is also hard to see how SGE or its effects could be commodified and turned into a product with paying customers.

Secondly, there is no policy champion for SGE among leading world politicians, in the way one might be able to identify champions for climate mitigation. The most likely champions for deploying SGE are likely to come from the authoritarian political right, such as the Trump administration. But they are currently focussed on climate inactivism (to put it generously), and many of Trump's followers are embroiled in the Chemtrail imaginary, and hostile to SGE.

Thirdly, like much climate politics, mobilising around the long-run is politically challenging. Whilst SGE may promise immediate cooling effects, its deployment and maintenance over a long period demands longer-term planning and consistency. SGE depends on an idea of emergency or urgency. But the climate condition is not widely perceived to be sufficiently serious or imminently threatening on the short timescales of the electoral and business cycles. This reduces the attractiveness of an unproven, possibly unreliable, technology.

Fourthly, capitalism is essentially reactive. It has little ability, as a system, to anticipate change or actively choose technological directions. In the drive to accumulate capital it generates social upheaval, technological disruption and environmental degradation, and then finds ways to adapt and reconstitute itself, in the process generating further upheaval, disruption and degradation. On this reasoning, SGE may be conceptually facilitative of capitalism's maintenance, but will not be pre-emptively embraced.

Perhaps the paradox is best explained by a double feature of SGE— that it is both stabilising *and* disruptive of the dominant order of "imperial market globalism" (Steger 2009). SGE does indeed have the character of a technical solution to an ecological problem and appears compatible with 'business-as-usual'. But it also seems to entail at least the possibility of substantial disruption of existing institutions, as my discussion of emergency and geo-politics above suggests. SGE might therefore be '*hyper-compatible*' with the dominant order: its compatibility is accompanied by disruptive characteristics. This can make SGE unattractive to those elements of the dominant order for whom the predictability and continuity of existing institutional structures and investment horizons is important, and disruption threatening. The more so because it is precisely the richest and most powerful inhabitants of the Earth who have the greatest capacity (by moving, cocooning and air conditioning) to put up with climate changes which exceed rises of two, three or even four degrees centigrade.

Values

Values, in the sense of who counts and what matters and how the world ought to be, are pervasive in contestation over SGE. The institutional reports and assessments of SGE rarely make their values explicit. Although the importance of ethics and values is widely acknowledged, consideration of these is typically narrowed to be about whether the possibility of deploying SGE will lead to decreased attention to mitigation (the so-called 'moral hazard' problem), and to discussion of governance questions. As one of the main contributors to the report of the IPCC Expert Group on Geoengineering, Scott Barrett, put it: "The central social, political, legal, and ethical challenges posed by this technology *all* concern governance" (IPCC 2012: 24, my emphasis).

Whilst human purpose and relational questions (which I shorthand as 'intrinsic values' questions) are sometimes acknowledged in the institutional reports, they are typically then ignored or ring-fenced into a short separate consideration of 'ethics'. Intrinsic values questions—such as 'are we (or scientists) entitled to remake the climate?', 'are there limits to what humans should do?', 'what sort of humans do we want to be?', 'are we entitled to commit future generations into having to maintain SGE?', 'isn't SGE un-natural?', 'what relationship with the more-than-human world do we aspire to?'—are glossed over, even though these are precisely the questions raised in public conversations and opinion studies and indeed by many climate scientists. The net effect is to focus on how SGE might be enabled (resolving values concerns through suitable governance mechanisms), rather than whether SGE is desirable or should be embraced at all.

The Ethics of SGE

There is now an extensive academic literature, and from a range of ethical perspectives, exploring the ethics of SGE (see for example Preston 2012; Gardiner 2011a; Corner and Pidgeon 2010; Di Paola 2013; Gardiner and Fragnière 2018). The issues explored are many and various. Questions of consent (and authority) arise repeatedly. Who gets to decide to embark upon, or even experiment, with this planet-changing technology? And, if embarked upon, who gets to set the global thermostat, or decide on termination of the intervention? Such questions are related to, but go deeper than, governance considerations. They are deeply bound up with *justice* concerns, such as whether the voice of the few overwhelms the voices of the many (see for example Macnaghten and Szerszynski 2013), or the imagined effects of SGE on the world's poorest. Justice concerns also have an inter-generational dimension. Given that SGE, once embarked upon, likely entails a multi-generational, even 'millennial', commitment, what obligations do current generations have towards future ones? (see for example Gardiner 2011a; Hordequin 2012; Preston 2016). Gardiner has suggested it is especially challenging to avoid "moral corruption"—the "subversion of our moral discourse to our own ends" (Gardiner 2011a)—when we consider questions such as these. Although Healey and Rayner argue that SGE raises no ethical questions not already raised in general considerations of climate change, they concede that the human-nature relationship is "underdeveloped in the mainstream climate change literature" (2015: 11–12).

Competing perspectives around intrinsic values are at the heart of disagreements around the desirability of SGE. Is the very idea of SGE and controlling the skies something both brazen and 'un-natural', a sign of hubris? Or have circumstances on Earth changed so much that it is necessary to embrace our allocated role as the 'God species' (Lynas 2011), "... consciously admitting that we live on a managed planet" (Keith quoted in Goodell 2010: 45)? Others, favourably disposed to SGE, argue that discussing the perfect ethical standpoint is to promote inaction: we are in a climate emergency, and so the most ethical geoengineering is 'whichever one we've got now" (Briggle 2018: 189). Others still, suggest we have to choose SGE for environmental reasons, as a 'less bad' solution. As Ken Caldeira, an active geoengineer and participant in some of the institutional reports, has put it: "Which is the more environmentally sensitive thing to do: let the Greenland ice sheet collapse and polar bears become extinct, or throw a little sulfate in the stratosphere?" (Caldeira 2007).

Such questions, rooted in contested values, in some respects reflect pre-existing arguments within environmental discourse (Dryzek 2005). But they do so in greatly sharpened form. As Clive Hamilton has put it: "what matters ethically about geoengineering is not only the outcome but also the human disposition it reveals" (Hamilton 2011: 18). Many of the ethical questions which geoengineering prompts are large ones and revolve around deeply-held values and visions regarding how the world is and how it should be. They are bound up in attitudes towards technology, and rooted in assumptions about nature, the human-nature relationship, and the permissible bounds of human action. There are a range of views on these issues within 'modern' Western cosmologies, even before coming to the range of views contained in competing ontological perspectives on the relationship between nature and culture (see Descola 2013 [2005]; Wong 2015). My purpose in this section is not to explore these ethical questions in detail, nor to understate their complexity. Rather, I want to understand how the more mainstream institutional assessments handle the ethical questions. Typically, as we will see, these acknowledge that there are ethical issues but then narrow and marginalise them.

Values and Ethics in the Institutional Reports

Two highly simplifying tropes are found repeatedly in the institutional reports. Firstly, that ethics is reduceable to moral hazard. Secondly, that ethical concerns are resolved by protocols and governance allowing the project to continue rather than halting it. The Royal Society report states that "the greatest challenges to the successful deployment of geoengineering may be the social, ethical, legal and political issues associated with governance" (2009: xi). It notes the view that geoengineering "may be intrinsically unethical" (2009: 46), but then does not engage with this troubling observation, thereby effectively discarding it. The report notes that there are different generic schools of philosophical thought (although only European traditions are considered). In practice it favours utilitarian/consequentialist approaches but acknowledges that these "probably tend towards a more favourable view" of geoengineering (p. 46). It treats the 'justice' questions as largely resolvable through the development of good governance frameworks (p. 46). Its overall discussion of ethics prioritises 'moral hazard': the term widely used in the geoengineering literature to express the idea that raising the possibility of SGE will detract from efforts to mitigate emissions. The report argues that "if it could be shown empirically that the moral hazard issue was not serious, one of the main ethical objections to geoengineering would be removed" (p. 39).[16] Not

---

[16] The widespread assumption has been that SGE would divert attention from mitigation and adaptation efforts and encourage 'business-as-usual' by making emissions reductions appear less urgent or necessary (see for example Hale 2012; Lin 2013). More recently, a substantial, largely speculative, literature has emerged which challenges these assumptions and argues, variously, that 'it would not' to 'it would encourage *more* mitigation' to 'if it reduces mitigation would that be a bad thing?'. For a survey of some of the arguments see Morrow (2014).

surprisingly, Gardiner has charged the Royal Society report with having a simplistic account of the ethical issues (Gardiner 2011b).

Throughout the NRC report (2015a) there is repeated reference to the phrase "social, political, legal and ethical issues", but only one page (out of 235) specifically exploring ethics. To the extent that there is a focus, it is on the specific and narrower question of research ethics. 'Moral hazard' is briefly touched upon. It is noted, in passing, that "[p]otential intergenerational implications compound the ethical issues regarding who has authority" to solar geoengineer (2015a: 135). The report mentions "potential psychological effects" of SGE and that there are debates over "the morality of deliberately taking control of the planet's temperature" (p. 135). The report concludes that "it is clear that further research on these ethical questions is required" (p. 135), although it seems far from clear that a lack of research is the problem. The instrumental nature of what is imagined becomes clearer in the subsequent sentence: "Research on the social implications and ecological and economic ramifications of deployment could better define if it is possible to mitigate societal concerns" (p. 135).

We can also examine the reports of the IPCC. Working Group 1 of the IPCC Fifth Assessment Report, which is generally unenthusiastic about SGE (or at least reflects the lack of agreement about its desirability), notes that "[t]here are also many (political, ethical, and practical) issues involving geoengineering that are beyond the scope of this report" (IPCC 2013: 632). The Working Group 2 portion of that report mentions, but does not explore, possible 'moral hazard' (IPCC 2014a: 1066). Chapter 3 of Working Group 3's report covers "social, economic, and ethical concepts and methods" (IPCC 2014b). Only six pages (out of 49) cover climate ethics, and one page within this focuses on geoengineering. This merely summarises the academic literature for and against geoengineering and notes that the 'moral hazard' argument is among the most prominent. The lead-up work for the Fifth Assessment Report, the expert group on geoengineering, acknowledges ethics but not altogether coherently. It argues that the central non-scientific challenges posed by SGE "all concern governance" (IPCC 2012: 24).

A similar ethics/governance conflation can be found in the report of the Bipartisan Policy Center. It notes that research on SGE may "raise new ethical, legal, and social issues of broad public concern" (BPC 2011:

19). It doesn't specify what these might be, but does recommend an advisory council to set "standards of oversight" (BPC 2011: 19), presumably of research oversight. The GAO study similarly suggests that "engagement with the public and U.S. decision-makers" may require "studies of economic, ethical, legal, and social issues and studies of systemic risks" (2011: vii). It cites, approvingly, a suggestion to bring "social scientists, ethicists, or trained risk assessors directly into laboratories to ensure early accounting for risks and social and ethical issues" (p. 52). Here ethics is treated as essentially risk management of potentially negative side-effects, to be resolved through suitable protocols.

When it comes to the question of what 'intrinsic values' or deeper ethical stances SGE assumes or requires, then this is generally either ignored or addressed perfunctorily. There are a few exceptions. The Umweltbundesamt (German Environmental Protection Agency) report, the assessment most sceptical of SGE (but least influential), notes that to gain societal acceptance involves asking "[w]hich ethical, moral, religious or aesthetic principles are touched on by application of this technology?" (Ginzky et al. 2011: 40). The SRMGI study, with its focus on governing research, lists "hubris and interference with nature" as one of eight potential concerns about SGE (2011: 22). The NOVIM study's framing of the ethical issues is unusually broad in scope, for the institutional literature. It cites the centrality of "beliefs about humanity's role in the natural world that are opposed to intentional human modification of the Earth's climate" (Blackstock et al. 2009: v). Almost alone in the institutional literature, it regards 'moral hazard' as a socio-political rather than an ethical issue. But, other than these observations, it sees ethical and socio-political issues as beyond the scope of its study. Indeed, none of these reports takes the intrinsic values issues much beyond the brief comments cited here, notwithstanding extensive public and academic reflection on these.

Overwhelmingly, the substantial ethical concerns which surround SGE are either glossed over or narrowed substantially in the institutional reports. Whilst these may note the big 'intrinsic values' questions as well as the intergenerational issues, or even acknowledge that SGE may be 'inherently unethical', they then largely ignore these observations. One consequence is that whilst ethical concerns may be conceded to be central, they are absent from the proposed criteria for evaluating geoengineering. For the Royal Society affordability, effectiveness, timeliness and safety are the key evaluation criteria (2009: 48). For the NRC (2015a) a similar, albeit more elaborate and complex, set of criteria is adopted. Criteria related to justice or consent concerns do not make an appearance, and neither do 'intrinsic values' concerns.

Ignoring and Narrowing Ethical Concerns

SGE raises major ethical questions, including about how humans are and should be in the world, and whether they have the right to re-mould the climate and the world so fundamentally, and whether SGE is 'inherently unethical'. But in the institutional reports such questions are persistently ignored or narrowed in three key ways. Firstly, intrinsic values issues are either glossed over, ignored or simply noted without further reflection, or treated as beyond the scope of the assessment.

Secondly, 'moral hazard' is treated as the most important ethical issue. It occupies centre stage in almost all the ethical sections of the institutional reports. Whilst the word 'moral' suggests that ethical questions are in play, it is hard to understand why the possibility of SGE leading to reduced mitigation should be treated as an ethical issue rather than a socio-political one, a question of climate policy choices. Further, what does it mean to centre ethical attention around a concept drawn from actuarial science and insurance economics? Reducing ethics to 'moral hazard' not only silences ethical concerns. It also, in effect, privileges a cost-benefit way of looking at the world and of containing ethical concerns within that framework. In so doing it works to normalise SGE.

Thirdly, the institutional literature typically reduces consent and justice questions, which revolve around ethical accountability, to governance questions. And governance is, in turn, as Gardiner notes, narrowly construed as institutional arrangements regarding which law or institution should take charge (2011b: 170). A plethora of institutional initiatives have tried to develop codes of conduct and suggested governance arrangements. This includes the Solar Radiation Management Governance Initiative (SRMGI 2011), the Asilomar initiative (ASOC 2010), the

Oxford Principles (see Rayner et al. 2013), and the Carnegie Climate Geoengineering Governance Initiative (C2G2 2018), among others.

As noted earlier in this Chapter, there is resistance among the most powerful to the adoption of globally inclusive governance frameworks. Perhaps not surprisingly then, governance questions are narrowed further and commonly end up focussing on research governance. In the institutional literature this is typically operationalised as the need for research protocols outlining the conditions under which experiments might take place or field testing be permitted, and debates as to whether any rules should be more or less voluntary. These, in turn, draw on the important, but narrow, field of laboratory and research ethics. Paul Nurse, the President of the Royal Society, has argued, for example, that geoengineering is akin to testing pharmaceuticals and should be similarly encouraged (2011). A more sceptical observer might note the absence, in this analogy, of the associated principle of 'informed consent' and how this might be applied when experimenting with the Earth itself!

In glossing over, ignoring or narrowing the ethical issues, it is hard not to conclude that the institutional literature largely inhabits the terrain of what Gardiner labels 'moral corruption'—the self-interested subversion of moral discourse.

* * *

The approach adopted to developing Knowledge about SGE follows the heavily scientised pattern of much existing climate policy. The scientific and the technical are elevated above other forms of knowing, and climate reductionism is the predominant approach used to project what the social impacts of SGE might be. The radical uncertainty which adheres to what is entailed by SGE is conceived as a standard risk and probability problem, resolvable if only more data could be gathered. And the utilitarian economic approaches embraced facilitate the calculation of the direct cost of SGE but not its societal and world-making costs. These ways of knowing act to normalise SGE as both potential technology and climate policy. They privilege particular expertise to legitimate a proposal that a small group of people should reshape our climate/s and remake our world/s.

The many Values and ethical questions which accompany any consideration of SGE are simultaneously acknowledged and narrowed in the institutional assessments. The widespread intrinsic values concerns, which have substantial popular resonance, which suspect that SGE may be 'inherently unethical' and undesirable, and crosses the boundary of what humans should do, are largely put to one side. The general assumption appears to be that having laid the scientific and technical foundation for assessment of SGE, the values questions can be handled by others, elsewhere and later, or are irrelevant in the face of needing to 'do something' about climate. These too are normalising moves.

The relationship to Power is more complex. The invocation of climate emergency is rhetorically powerful and helps make radical intervention in the planet thinkable, even as radical intervention in the economic order driving climate change is passed over in silence because it is unthinkable. But this normalising move is balanced by an acknowledgement that SGE has the potential to exacerbate significant geo-political instability.

In short, in the institutional assessments, the ways in which Knowledge, Values and Power are understood helps to normalise SGE and to reinforce the Imperial imaginary of what SGE is and should be. This is true even where these reports do not explicitly support SGE's embrace and potential deployment. However, it is important to note that just because the assessments largely work to normalise SGE they have not yet succeeded in doing so. Oppositional imaginaries retain substantial popular purchase. Climate scientists and climate policymakers remain largely sceptical of SGE. There is widespread recognition that debates around SGE are about more than the utility of a mundane technology to address climate change. They are about the futures we imagine and the worlds in which we aspire to live.

References

Arctic Methane Emergency Group. (2014, December 4). Press release. Retrieved January 20, 2019, from https://web.archive.org/web/20141216194538/ameg.me/

Armeni, C., & Redgwell, C. (2015). International legal and regulatory issues of climate geoengineering governance: Rethinking the approach. *Climate Geoengineering Governance Working Paper Series*: 21. Retrieved January 9, 2019, from http://www.geoengineering-governance-research.org/perch/resources/workingpaper21armeniredgwelltheinternationalcontextrevise-.pdf

Asilomar Scientific Organizing Committee (ASOC). (2010). *The Asilomar conference recommendations on principles for research into climate engineering techniques*. Washington, DC: Climate Institute.

Barrett, S. (2008). The incredible economics of geoengineering. *Environmental and Resource Economics, 39*(1), 45–54.

Barry, A., Born, G., & Weszkalnys, G. (2008). Logics of interdisciplinarity. *Economy & Society, 37*(1), 20–49.

Battisti, D. S., & Naylor, R. L. (2009). Historical warnings of future food insecurity with unprecedented seasonal heat. *Science, 323*(5911), 240–244.

Beck, U. (2006). Living in the world risk society. *Economy and Society, 35*(3), 329–345.

Bellamy, R., Chilvers, J., Vaughan, N. E., & Lenton, T. M. (2012). A review of climate geoengineering appraisals. *WIREs Climate Change, 3*(6), 597–615.

Bellamy, R., Chilvers, J., Vaughan, N. E., & Lenton, T. M. (2013). 'Opening up' geoengineering appraisal: Multi-criteria mapping of options for tackling climate change. *Global Environmental Change, 23*, 926–937.

Bickel, J. E., & Agrawal, S. (2013). Reexamining the economics of aerosol geoengineering. *Climatic Change, 119*(3–4), 993–1006.

Bickel, J. E., & Lane, L. (2009). *An analysis of climate engineering as a response to climate change*. Copenhagen: Copenhagen Consensus Center.

Bickel, J. E., & Lane, L. (2012). *Challenge paper: Climate change, climate engineering R&D*. Copenhagen: Copenhagen Consensus Center.

Bipartisan Policy Center (BPC). (2011). *Geoengineering: A national strategic plan for research on the potential effectiveness, feasibility, and consequences of climate remediation technologies*. Washington, DC: Bipartisan Policy Center Task Force on Climate Remediation Research.

Blackstock, J. J., Battisti, D. S., Caldeira, K., Eardley, D. M., Katz, J. I., Keith, D. W., et al. (2009). *Climate engineering responses to climate emergencies*. Santa Barbara: Novim. Retrieved January 9, 2019, from http://arxiv.org/pdf/0907.5140

Brennan, J. (2016). CIA director on the geopolitical risks of climate geoengineering. Video and transcript of talk to Council on Foreign Relations, July 2016. Retrieved January 12, 2019, from https://climateandsecurity.org/2016/07/25/cia-director-on-the-geopolitical-risks-of-climate-geoengineering/#more-9259

Briggle, A. (2018). Beware of the toll keepers: The ethics of geoengineering ethics. *Ethics, Policy & Environment, 21*(2), 187–189.

C2G2 (Carnegie Climate Geoengineering Governance Initiative). (2018, November). *Governing Solar Radiation Modification (SRM)*. Retrieved January 9, 2019, from https://www.c2g2.net/wp-content/uploads/C2G2_Solar-Brief-hyperlink.pdf

Caldeira, K. (2007, October 24). How to cool the globe. *New York Times*. Retrieved January 12, 2019, from http://www.nytimes.com/2007/10/24/opinion/24caldiera.html?_r=0

Calhoun, C. (2010). The idea of emergency: Humanitarian action and global (dis)order. In D. Fassin & M. Pandolfi (Eds.), *Contemporary states of emergency: The politics of military and humanitarian interventions* (pp. 29–58). New York: Zone Books.

Castree, N. (2015). Geography and global science: Relationships necessary, absent, and possible. *Geographical Research, 53*(1), 1–15.

Castree, N., Demeritt, D., & Liverman, D. (2009). Introduction: Making sense of environmental geography. In N. Castree, D. Demeritt, D. Liverman, & B. Rhoads (Eds.), *A companion to environmental geography*. Oxford: Blackwell.

Corner, A., & Pidgeon, N. (2010). Geoengineering the climate: The social and ethical implications. *Environment, 52*(1), 24–37.

Curry, J. A., & Webster, P. J. (2011). Climate science and the uncertainty monster. *Bulletin of the American Meteorological Society, 92*, 1667–1682.

Curvelo, P. (2012). Exploring the ethics of geoengineering through images. *The International Journal of the Image, 2*(2), 177–198.

Descola, P. (2013 [2005]). *Beyond nature and culture*. Translated from French (2005) by J. Lloyd. Chicago, IL: University of Chicago Press.

Di Paola, M. (2013). Climate change and moral corruption. Symposium: A changing moral climate. *Philosophy and Public Issues* (New Series), Special issue, *3*(1), 55–67.

Doppelt, B. (2012). *The power of sustainable thinking: How to create a positive future for the climate, the planet, your organization and your life*. London: Earthscan.

Dryzek, J. S. (2005). *The politics of the Earth: Environmental discourses* (2nd ed.). New York: Oxford University Press.

Engelke, P., & Chiu, D. (2016). *Climate Change and US National Security: Past, present, future*. Washington, DC: Atlantic Council.

ETC Group. (2010, November). *Geopiracy: The case against geoengineering* (2nd ed.). Retrieved January 9, 2019, from https://www.cbd.int/doc/emerging-issues/etcgroup-geopiracy-2011-013-en.pdf

ETC Group/Biofuelwatch. (2017). *The Big Bad Fix: The case against climate geoengineering*. Retrieved January 9, 2019, from http://etcgroup.org/sites/www.etcgroup.org/files/files/etc_bbf_mar2018_us_v1_web.pdf

Funtowicz, S. O., & Ravetz, J. R. (1993). Science for the post-normal age. *Futures, 25*(7), 735–755.

GAO (Government Accountability Office). (2010). *Climate change: A coordinated strategy could focus federal geoengineering research and inform governance efforts*. GAO-10-903. Washington, DC: U.S. Government Accountability Office. Retrieved January 9, 2019, from https://www.gao.gov/assets/320/310105.pdf

GAO (Government Accountability Office). (2011). *Climate engineering: Technical status, future directions, and potential responses*. GAO-11-71. Washington, DC: U.S. Government Accountability Office. Retrieved January 9, 2019, from http://www.gao.gov/new.items/d1171.pdf

Gardiner, S. (2011a). *A perfect moral storm: The ethical tragedy of climate change*. Oxford: Oxford University Press.

Gardiner, S. M. (2011b). Some early ethics of geoengineering the climate: A commentary on the values of the royal society report. *Environmental Values, 20*(2), 163–188.

Gardiner, S. M. (2013). Why geoengineering is not a 'global public good', and why it is ethically misleading to frame it as one. *Climatic Change, 121*(3), 513–525.

Gardiner, S. (2014). Why 'global public good' is a treacherous term, especially for geoengineering. *Climatic Change, 123*(2), 101–106.

Gardiner, S. M., & Fragnière, A. (2018). The tollgate principles for the governance of geoengineering: Moving beyond the Oxford principles to an ethically more robust approach. *Ethics, Policy & Environment, 21*(2), 143–174.

Ginzky, H., Herrmann, F., Kartschall, K., Leujak, W., Lipsius, K., Mäder, C., et al. (2011). *Geoengineering: Effective climate protection or megalomania?* Dessau-Roßlau: Umweltbundesamt.

Goes, M., Tuana, N., & Keller, K. (2011). The economics (or lack thereof) of aerosol geoengineering. *Climatic Change, 109*(3–4), 719–744.

Goodell, J. (2010). *How to cool the planet: Geoengineering and the audacious quest to fix Earth's climate*. Melbourne: Scribe.

Hale, E. (2012, May 16). Geoengineering experiment cancelled due to perceived conflict of interest. *The Guardian*. Retrieved January 12, 2019, from https://www.theguardian.com/environment/2012/may/16/geoengineering-experiment-cancelled

Hamilton, C. (2011). *Ethical anxieties about geoengineering*. Paper presented to a conference of the Australian Academy of Science Canberra, 27 September. Retrieved January 12, 2019, from http://clivehamilton.com/ethical-anxieties-about-geoengineering/

Hamilton, C. (2013). *Earthmasters: Playing God with the climate*. Crow's Nest, NSW: Allen & Unwin.

Hansson, A. (2014). Ambivalence in calculating the future: The case of re engineering the world. *Journal of Integrative Environmental Sciences, 11*(2), 125–142.

Healey, P., & Rayner, S. (2015). Key findings from the Climate Geoengineering Governance (CGG) project. *Climate Geoengineering Governance Working Paper Series*: 25. Retrieved January 9, 2019, from http://www.geoengineering-governance-research.org/perch/resources/workingpaper25healeyraynerkeyfindings-1.pdf

Heyward, C., & Rayner, S. (2013). A curious asymmetry: Social science expertise and geoengineering. *Climate Geoengineering Governance Working Paper Series*: 007. Retrieved January 9, 2019, from http://geoengineering-governance-research.org/perch/resources/workingpaper7heywardrayneracuriousasymmetry.pdf

Hordequin, M. (2012). Justice, recognition, and climate geoengineering. In C. J. Preston (Ed.), *Engineering the climate: The ethics of Solar Radiation Management*. Lanham, MD: Lexington Books.

Horton, J. B. (2015). The emergency framing of solar geoengineering: Time for a different approach. *The Anthropocene Review, 2*(2), 147–151.

Horton, J. B., Keith, D. W., & Honegger, M. (2016). Implications of the Paris agreement for carbon dioxide removal and solar geoengineering. Harvard Project on Climate Agreements viewpoint paper, July. Retrieved January 9, 2019, from https://www.belfercenter.org/sites/default/files/files/publication/160700_horton-keith-honegger_vp2.pdf

Hulme, M. (2011). Reducing the future to climate: A story of climate determinism and reductionism. *Osiris, 26*, 245–266.

Hulme, M. (2014). *Can science fix climate change? A case against climate engineering*. Cambridge, UK: Polity Press.

IPCC (Intergovernmental Panel on Climate Change). (2007). Climate Change 2007: Synthesis report. In R. K. Pachauri & A. Reisinger (Eds.), *Contribution of Working Groups I, II, and III to the Fourth Assessment Report of the Intergovernmental Panel on Climate Change*. Geneva: IPCC.

IPCC (Intergovernmental Panel on Climate Change). (2012). *Meeting report of the Intergovernmental Panel on Climate Change expert meeting on geoengineering*

(O. Edenhofer, R. Pichs-Madruga, Y. Sokona, C. Field, V. Barros, T. F. Stocker, Q. Dahe, J. Minx, K. Mach, G.-K. Plattner, S. Schlömer, G. Hansen, & M. Mastrandrea, Eds.). Potsdam: IPCC Working Group III Technical Support Unit, Potsdam Institute for Climate Impact Research. Geneva: IPCC.

IPCC (Intergovernmental Panel on Climate Change). (2013). Climate Change 2013: The physical science basis. In *Contribution of Working Group I to the Fifth Assessment Report of the Intergovernmental Panel on Climate Change*. Cambridge, UK and New York: Cambridge University Press.

IPCC (Intergovernmental Panel on Climate Change). (2014a). Climate Change 2014: Impacts, adaptation, and vulnerability. In *Contribution of Working Group II to the Fifth Assessment Report of the Intergovernmental Panel on Climate Change*. Cambridge, UK and New York: Cambridge University Press.

IPCC (Intergovernmental Panel on Climate Change). (2014b). Climate Change 2014: Mitigation of climate change. In *Contribution of Working Group III to the Fifth Assessment Report of the Intergovernmental Panel on Climate Change*. Cambridge, UK and New York: Cambridge University Press.

Jacobson, M. Z., Delucchi, M. A., Bauer, Z. A., Goodman, S. C., Chapman, W. E., Cameron, M. A., et al. (2017). 100% clean and renewable wind, water, and sunlight all-sector energy roadmaps for 139 countries of the world. *Joule, 1*(1), 108–121.

Jasanoff, S. (2004). The idiom of co-production. In S. Jasanoff (Ed.), *States of knowledge: The co-production of science and social order*. London and New York: Routledge.

Keith, D. W. (2013). *A case for climate engineering*. Cambridge, MA: MIT Press.

Klein, N. (2015). *This changes everything: Capitalism vs. the climate*. Melbourne: Penguin Books.

Krauss, W., Schäfer, M. S., & Von Storch, H. (2012). Introduction: Post-normal climate science. *Nature and Culture, 7*(2), 121–132.

Lagorio, J. J. (2007, November 10). U.N.'s Ban says global warming is "an emergency". *Reuters*. Retrieved January 28, 2019, from https://www.reuters.com/article/environment-antarctica-un-ban-dc-idUSN0923477720071110

Lane, L., & Bickel, J. E. (2013). *Solar Radiation Management: An evolving climate policy option*. Washington, DC: American Enterprise Institute.

Lane, L., Caldeira, K., Chatfield, R., & Langhoff, S. (2007). *Workshop report on managing solar radiation*. NASA Ames Research Centre & Carnegie Institute of Washington, Moffett Field, CA, 18–19 November. Hanover, MD: NASA. (NASA/CP-2007-214558).

Lempert, R. J., & Prosnitz, D. (2011). *Governing geoengineering research: A political and technical vulnerability analysis of potential near-term options*. Santa Monica, CA: RAND Corporation.

Lenton, T. M. (2012). Arctic climate tipping points. *Ambio, 41*(1), 10–22.
Lenton, T. M. (2013). Can emergency geoengineering really prevent climate tipping points? *Geoengineering our climate: Working paper and opinion article series.* Retrieved January 9, 2019, from https://geoengineeringourclimate.wordpress.com/2013/06/25/can-emergency-geoengineering-really-prevent-climate-tipping-points-opinion-article/
Lin, A. C. (2013). Does geoengineering present a moral hazard? *Ecology Law Quarterly, 40*(3), 673–712.
Lynas, M. (2011). *The god species: How the planet can survive the age of humans.* London: Fourth Estate.
MacKenzie, D. (2009). Making things the same: Gases, emission rights and the politics of carbon markets. *Accounting, Organizations and Society, 34*, 440–455.
MacKerron, G. (2014). Costs and economics of geoengineering. *Climate Geoengineering Governance Working Paper Series*: 013. Retrieved January 9, 2019, from http://www.geoengineering-governance-research.org/perch/resources/workingpaper13mackerroncostsandeconomicsofgeoengineering.pdf
MacMartin, D. G., Caldeira, K., & Keith, D. W. (2014). Solar geoengineering to limit the rate of temperature change. *Philosophical Transactions of the Royal Society A, 372*(2031), 1–13.
Macnaghten, P., & Szerszynski, B. (2013). Living the global social experiment: An analysis of public discourse on Solar Radiation Management and its implications for governance. *Global Environmental Change, 23*, 465–474.
Markusson, N., Ginn, F., Ghaleigh, N. S., & Scott, V. (2014). 'In case of emergency press here': Framing geoengineering as a response to dangerous climate change. *WIREs Climate Change, 5*(2), 281–290.
McClellan, J., Keith, D., & Apt, J. (2012). Cost analysis of stratospheric albedo modification delivery systems. *Environmental Research Letters, 7*(3), 1–8.
McCusker, K. E., Battisti, D. S., & Bitz, C. M. (2012). The climate response to stratospheric sulfate injections and implications for addressing climate emergencies. *Journal of Climate, 25*(9), 3096–3116.
Morrow, D. R. (2014). Why geoengineering is a public good, even if it is bad. *Climatic Change, 123*(2), 95–100.
National Research Council (NRC). (2015a). *Climate intervention: Reflecting sunlight to cool Earth.* Washington, DC: National Academy of Sciences.
National Research Council (NRC). (2015b). *Climate intervention: Summary report.* Washington, DC: National Academy of Sciences.

Nerlich, B., & Jaspal, R. (2012). Metaphors we die by? Geoengineering, metaphors, and the argument from catastrophe. *Metaphor and Symbol, 27*(2), 131–147.
Nurse, P. (2011, September 8). I hope we never need geoengineering, but we must research it. *The Guardian*. Retrieved January 9, 2019, from https://www.theguardian.com/environment/2011/sep/08/geoengineering-research-royal-society
Pinch, T., & Bijker, W. (1987). The social construction of facts and artifacts: Or how the sociology of science and the sociology of technology might benefit each other. In W. Bijker, T. P. Hughes, & T. Pinch (Eds.), *The social construction of technological systems: New directions in the sociology and history of technology* (pp. 17–44). Cambridge, MA: MIT Press.
Porter, T. M. (1995). *Trust in numbers: The pursuit of objectivity in science and public life*. Princeton, NJ: Princeton University Press.
Preston, C. J. (Ed.). (2012). *Engineering the climate: The ethics of Solar Radiation Management*. Plymouth, UK and Lanham, MD: Lexington Books.
Preston, C. (Ed.). (2016). *Climate justice and geoengineering ethics and policy in the atmospheric anthropocene*. London: Rowman & Littlefield.
Rayner, S. (2014). To know or not to know? A note on ignorance as a rhetorical resource in geoengineering debates. *Climate Geoengineering Governance Working Paper Series*: 010. Retrieved January 9, 2019, from http://geoengineering-governance-research.org/perch/resources/workingpaper-10raynertoknowornottoknow-1.pdf
Rayner, S., Heyward, C., Kruger, T., Pidgeon, N., Redgwell, C., & Savulescu, J. (2013). The Oxford principles. *Climatic Change, 121*(3), 499–512.
Ricke, K., Morgan, M. G., Apt, J., Victor, D., & Steinbruner, J. (2008, May 5). *Unilateral geoengineering: Non-technical briefing notes for a workshop at the Council on Foreign Relations*, Washington, DC. Retrieved January 9, 2019, from http://www.cfr.org/content/thinktank/GeoEng_Jan2709.pdf
Rickels, W., Klepper, G., Dovern, J., Betz, G., Brachatzek, N., Cacean, S., et al. (2011). *Large-scale intentional interventions into the climate system? Assessing the climate engineering debate*. Scoping report conducted on behalf of the German Federal Ministry of Education and Research (BMBF), Kiel Earth Institute.
Royal Society. (2009). *Geoengineering the climate: Science, governance and uncertainty*. RS Policy document 10/09. London: Royal Society. Retrieved January 9, 2019, from https://royalsociety.org/~/media/Royal_Society_Content/policy/publications/2009/8693.pdf

Scott, K. (2013). International law in the anthropocene: Responding to the geoengineering challenge. *Michigan Journal of International Law, 34*, 309–358.
Sillmann, J., Lenton, T. M., Levermann, A., Ott, K., Hulme, M., Benduhn, F., et al. (2015). Climate emergencies do not justify engineering the climate. *Nature Climate Change, 5*, 290–292.
Smith, W., & Wagner, G. (2018). Stratospheric aerosol injection tactics and costs in the first 15 years of deployment. *Environmental Research Letters, 13*(12), 4001.
Solar Radiation Management Governance Initiative (SRMGI). (2011). *Solar Radiation Management: The governance of research*. Issued by the Environmental Defense Fund, The Royal Society, and The World Academy of Sciences.
Steger, M. B. (2009). *Globalisms: The great ideological struggle of the twenty-first century* (3rd ed.). Lanham, MD: Rowman & Littlefield.
Stern, N. (2007). *The economics of climate change: The Stern review*. Cambridge: Cambridge University Press.
Stilgoe, J. (2015). *Experiment earth: Responsible innovation in geoengineering*. Abingdon: Routledge.
Stirling, A. (2010). Keep it complex. *Nature, 468*, 23–30.
Surprise, K. (2018). Preempting the second contradiction: Solar geoengineering as spatiotemporal fix. *Annals of the American Association of Geographers, 108*(5), 1228–1244.
Szerszynski, B., & Galarraga, M. (2013). Geoengineering knowledge: Interdisciplinarity and the shaping of climate engineering research. *Environment and Planning A, 45*(12), 2817–2824.
United Nations. (1992). *Agenda 21*. United Nations Conference on Environment and Development, Rio de Janeiro, Brazil, 3–14 June. Retrieved January 9, 2019, from https://sustainabledevelopment.un.org/content/documents/Agenda21.pdf
UNFCCC (United Nations Framework Convention on Climate Change). (2015). *Adoption of the Paris agreement*. FCCC/CP/2015/L.9/Rev.1. Conference of the Parties, Paris, 12 December.
US House of Representatives Committee on Science and Technology (USHCST). (2010). *Engineering the climate: Research needs and strategies for international coordination*. Report by Bart Gordon.
Van Hemert, M. (2017). Speculative promise as a driver in climate engineering research: The case of Paul Crutzen's back-of-the-envelope calculation on solar dimming with sulfate aerosols. *Futures, 92*, 80–89.

Victor, D. G. (2011). *Global warming gridlock: Creating more effective strategies for protecting the planet.* Cambridge: Cambridge University Press.

Virgoe, J. (2009). International governance of a possible geoengineering intervention to combat climate change. *Climatic Change, 95,* 103–119.

Wegner, G., & Pascual, U. (2011). Cost-benefit analysis in the context of ecosystem services for human well-being: A multidisciplinary critique. *Global Environmental Change, 21*(2), 492–504.

Whiteside, K. H. (2006). *Precautionary politics: Principle and practice in confronting environmental risk.* Cambridge, MA: MIT Press.

Wong, P.-H. (2015). Confucian environmental ethics, climate engineering, and the "playing god" argument. *Zygon, 50*(1), 28–41.

Wood, G. D. (2014). *Tambora: The eruption that changed the world.* Princeton, NJ: Princeton University Press.

World Bank. (2012). *Inclusive green growth: The pathway to sustainable development.* Washington, DC: International Bank for Reconstruction and Development/World Bank.

Wynne, B. (1992). Uncertainty and environmental learning. *Global Environmental Change, 2*(June), 111–127.

6

Future Imaginings

We have seen that consideration of solar geoengineering (SGE) is no longer taboo, but that it has also not become normalised as an acceptable addition to the portfolio of climate policies and technologies. The institutional assessments, as I have shown in my analysis of knowledge, values and power in Chapter 5, have adopted a number of approaches which might have been expected to help normalise SGE. But they have typically withheld explicit endorsement of the technology and have not recommended its acceptance as a respectable third leg of climate policy.[1] We have seen too, in Chapter 4, that a number of competing imaginaries swirl around SGE. The futures they imagine are generally gloomy and dystopian.

What is the future of solar geoengineering? What is constraining its emergence as a 'normal' and deployable component of climate policy? Because the future is unpredictable this Chapter will inevitably be more speculative than previous Chapters. I will focus on both the ideational and the material and suggest that three variables will be critical. Firstly, changes to the climate of climate change. Most accounts suggest that what happens physically to the climate is the most critical factor in

---

[1] Experimental, relatively limited deployment of SGE, as recommended by SGE's chief proponents, has to date failed to win 'official' endorsement (NRC 2015)—although this has not stopped preparations for such experiments from proceeding.

whether or not SGE is deployed. I emphasise, rather, that the sociopolitical unfolding of climate change is key, and the lived experience of weather in place. Secondly, the extent to which existing powerful elites regard SGE as essential to stabilise, or re-stabilise, the dominant geopolitical and socio-economic order. Whilst SGE holds the promise of facilitating the ongoing reproduction of existing systems of production and consumption and existing relations of inequality and power, it may also undermine these very systems. Thirdly, whether the overwhelmingly dystopian vision of SGE which currently exists is replaced by a more positive vision.

As a not-yet-operational technology, SGE is already being fashioned by the many desires, beliefs and expectations which society projects onto it. How the competing imaginaries of SGE evolve will affect whether and, if so, how it is normalised and deployed.[2] When examining these I focus on, and critically discuss, two paradigms which are becoming evident in the Imperial framing and envisioning of SGE. The first is 'developmentalism' and draws on the dominant contemporary account of how the project of modernisation and globalisation should be achieved. The second is the novel, and increasingly ubiquitous, concept of the 'Anthropocene': that humanity is now the dominant geological force shaping the Earth. I conclude by suggesting that if SGE does materialise as an operational technology, it will do so as a sociotechnical imaginary of the Anthropocene.

The Climate of Power

Scientific analysis tells us CO_2 emissions and global average temperatures are still rising, and projected to rise further. The drivers are mainly anthropogenic. Rising greenhouse gas (GHG) concentrations are heavily driven by carbon-intensive economic, energy and transport systems, and by changes in land use such as de-forestation and more intensive agricultural practices (IPCC 2013). In short, rising GHG emissions are a function of a particular form of industrial modernity.

---

[2] Of course, the ideational does not exist as 'pure idea' or free-floating concept and cannot be understood only in the abstract. Rather, the ideational and the material are intertwined, as the term 'sociotechnical' implies.

The Paris climate talks of December 2015 set the goal of keeping warming below 2 °C compared to pre-industrial levels, whilst also including the aim to avoid exceeding 1.5 °C (UNFCCC 2015). This is generally agreed to be unachievable on the basis of existing national climate targets (Clémençon 2016). How much warming is likely to occur is dependent on many factors, including the extent to which ongoing GHG emissions are curtailed and CO_2 concentrations kept within touching distance of 400 ppm.

Most troublingly for climate policy-making, predictions of how the climate will behave are necessarily uncertain, and the 'liveliness' and non-linearity of the climate system means that the most reliable estimates are necessarily probabilistic, and typically cover a wide range of probabilities. Wagner and Weitzman, for example, show that whilst median estimates are of a temperature increase of 2.5 °C above pre-industrial associated with 550 ppm, the estimates also include so-called 'fat tails'—lower, but still troubling, probabilities of much more extreme temperature rises, in this case increases of greater than 6 °C (Wagner and Weitzman 2015: 53).

The societal effects of increases in temperature are also difficult to predict, and become increasingly difficult the higher above the present it is. The IPCC has, at various times, attempted to estimate the likely effects of various temperature increases: using its 'burning embers' image to illustrate this (IPCC 2001). A number of other studies have attempted to describe what the future may hold (for example Lynas 2008). In all cases their conclusions are *extremely* disturbing. Even with a rise of 1.5 °C, which is imminent, the effects are troubling (IPCC 2018). Coral reef ecosystems, for example, become largely unviable approaching 2 °C. A report prepared for the World Bank, specifically examining sub-Saharan Africa, South Asia and South-East Asia, expects that between 2 °C and 4 °C extreme climate events could push households into poverty traps, and affect crop yields and food security adversely (World Bank 2013). For Lynas, at 5 °C the planet becomes unrecognisable and billions can be expected to die (2008). Importantly, much depends on the speed of warming and the adaptability of societies to such changes. Climate change is no longer comprehensible as simply a backdrop or a cause that produces known effects. The reality is more complex and causation is multi-directional. Climate is both shaped by and actively shapes human activity.

How the warming trends unfold will, of course, impact on SGE's emergence (or not) as a potential climate policy and technology. But only in the most general terms can one conclude that the greater the global warming, the more attractive SGE becomes as a solution: three major caveats are needed.

Firstly, at the human level climate is experienced as weather. Average global temperature figures conceal as much as they reveal. For example, February 2016 was depicted as an exceptionally warm month in the Northern hemisphere winter, and global average temperatures were 1.35 °C above the 1951–1980 thirty year average, and 1.63 °C above pre-industrial (NASA GISS 2016). But the actual temperature changes in particular communities varied dramatically—a super-heated climate in the Arctic and across swathes of Russia; a large rise of up to 4 degrees in much of central Asia, Eastern Europe, the Middle East, and the forest-dense regions of central South America and central Africa; and even cooling in some places. Climate change is not experienced as a global average, but rather as particular lived conditions in particular societies and communities, each with varied climate sensitivity and varied ability, resources and capacity for adapting to such changes. In some cases, warmer winters may even be embraced by those living in higher latitudes (Nyvold 2015; but see Buck 2018).

Secondly, for considerations of SGE, the rate of change is likely to be more significant, socially and politically, than the magnitude of temperature change. A slow, albeit steady, rise in temperatures—as has occurred over the past decades—fails to prompt the necessary sense of urgency which seems to be a pre-requisite for SGE to be embraced. Gradual temperature increases are less perceptible to humans, and are often adapted to as the 'new normal'. There are still human casualties of course, not to mention casualties across ecosystems and among other species. But without an 'event' we are left with slow climate disruption. This is analogous to what Rob Nixon has called, in another context, "slow violence": "violence that occurs gradually and out of sight, a violence of delayed destruction that is dispersed across time and space, an attritional violence that is typically not viewed as violence at all" (2011: 2). Only major climate events, such as major extreme weather events, impacting significant numbers of people and capable of being visually communicated, are likely to prompt calls for the deployment of SGE. Typhoon Haiyan, which struck the Philippines in

2013 and claimed an estimated 6300 lives with over 1000 missing, is perhaps an example, although, since more remote areas and poorer residents of the country bore the brunt of this extreme weather event, it lacked global tele-visibility (Lagmay et al. 2015). The Japanese tsunami of 2011 is perhaps a better example of what I have in mind as a trigger event, since it prompted a major about turn in that country's nuclear energy policy, although it is, of course, not a climate change example.

Thirdly, and following on from the above, any such climate event will lead to significant calls for SGE only to the extent that it impacts people, places and societies that 'matter', as seen by those with the power and ability to embark on SGE, not to mention the political will. Above all this means as seen by the United States government and/or its key allies. To this extent, Victor's "hard-nosed realist" approach, as discussed in Chapter 4, is surely relevant. In short, SGE is unlikely to be embarked upon to save the polar bear or the sparsely-populated Arctic circle.[3] And many consecutive days of temperatures close to 50 °C in India, as occurred in 2016, will carry considerably less weight than a similar event in California would when it comes to legitimating SGE as a response.

In short, climate developments matter, and a warming climate makes the deployment of SGE more likely. But much depends on whether such developments occur and are interpreted as major climatic events, and whether these impact heavily on the inhabitants of countries with the geo-political power to embark on SGE. The power of climate and the climate of power are entwined, and this is especially evident in considerations around SGE.

A Technology of the Powerful

Without reprising long-standing debates on technological determinism in the field of Science and Technology Studies, it is evident that technologies can have politics, or at least be associated with political effects, even when not explicitly designed to do so. Their very design may tend to

---
[3] Some scientific studies have attempted to model how SGE might be used to re-freeze the Arctic and expand the extent of sea-ice (Jackson et al. 2015). But at the same time there are powerful interests in favour of using declining sea-ice cover to open cheaper shipping routes and conduct oil drilling.

reinforce centralised rather than decentralised authority (think solar power vs. nuclear power). They may tend to reinforce rather than undermine existing power relations and the interests of the powerful (think nuclear weapons).

SGE is explicitly aimed at intentionally reversing or slowing down global warming. In so doing it inaugurates novel climate/s and re-shapes global climate in regionally varied and largely uncertain ways, and with effects on the Earth's human and more-than-human occupants, and on the societies and biomes they constitute and inhabit. SGE is manifestly a powerful technology. Less commonly acknowledged is that it is also a technology of the powerful.

SGE enables, or at the very least is compatible with, the continuation of existing, and otherwise environmentally unsustainable, patterns of extraction, production, consumption and trade. All the competing imaginaries outlined in Chapter 4 acknowledge this, even as they adopt different stances regarding the desirability of this compatibility. The three strands of the Imperial imaginary regard this compatibility as, variously, highly desirable (Market), necessary for global stability (Geo-management), or inevitable, absent alternatives (Salvation).

Embarking on SGE is a long-term commitment that could range from decades (the short estimate) to centuries. How long is a source of disagreement. Beyond the most limited aerosol dosages, the consensus appears to be that it cannot be terminated quickly or easily without risking the so-called 'termination effect', an angry bounce-back of previously constrained warming. Responsible deployment requires ongoing aerosol injections and therefore stable and predictable governance over periods exceeding the life of any existing government or global institution and most socio-economic orders in human history to date. This assumes the maintenance, or at least stability, of existing structures/regimes of global power.

Relatedly, SGE would appear to be inimical to democracy, thought of as any system which allows the demos some meaningful ability to constrain the actions of the elite and replace incumbent rulers. Democracy implies the possibility of reversibility, accountability and recall, and typically operates on timescales incompatible with SGE. Further, SGE imposes climatic effects on everyone. It is difficult to conceive of circumstances in which its deployment might be multilaterally agreed by a wide range of countries and people. Nor could non-consenting states withdraw from the climatic consequences of SGE. Even if there was a meaningful level of multilateral agreement around commencing SGE, the governance of the management and ongoing implementation of deployed SGE would ideally require multi-decadal stability and consistency. Autocracy rather than democracy.

Dilemmas of this kind are implicitly acknowledged in the Imperial imaginary with its vision that a single global power or 'club' of powers might agree to deploy SGE. Anxiety about agreement can also be seen in efforts by some in the global North to find a legitimate (and legitimising?) governance structure for SGE, whilst at the same time resisting efforts to prohibit SGE under existing multilateral global institutions and treaties. A technology that is not widely accepted can only be imposed, and this is necessarily an exercise of coercive power.[4]

---

[4] I am arguing that SGE is 'inimical' rather than 'incompatible' with broad consent for its deployment. As this book was going to press, in a critique of the democratic incompatibility argument, Horton et al. (2018) argue that SGE could be governed in either a democratic or authoritarian manner and that there is nothing in the technology that precludes democratic governance. Proponents of the argument that SGE is incompatible with democracy are charged with technological determinism, and of relying on a particular (deliberative) notion of what democracy is. It is suggested these proponents have a naïve and utopian view of what constitutes democracy, and that democracy does not preclude centralised control, nor does it mean citizens can opt out of decisions once made. My own argument does not rely on a naïve or deliberative view of what democracy is. Nor do I accept that the inability of individual citizens to opt-out of democratic decisions can be extended to the international sphere: indeed, in the multilateral sphere opt-out remains the norm (the US opting-out of the UNFCCC is a case in point). Further, whilst the theoretical point about technological determinism is taken, it may be that SGE has some inherent features which make it an exceptional case—flowing from the time horizons, the difficulty of termination, and the non-containable weather effects which determinedly flow from it.

Horton et al. also assert that existing global politics and its institutions, norms and expectations are "more or less democratic" (2018: 10). This is a view many from the global South would find hard to agree with, even if the observation is only confined to inter-state relations, since it ignores the substantial power imbalances in the international domain.

The future is inevitably speculative. It may, in theory, be the case that SGE will be deployed and governed multilaterally in a sufficiently consensual way to be accepted as legitimate and democratic. But it is hard to see much evidence for this becoming the case. Indeed, the United States today is not only the major locus for SGE work and the likely driver of any deployment. It also currently displays a willingness to disrupt any semblance of rules-based 'more or less democratic' order and predictability, in relation to climate change and much else, which is deemed to be against its interests. With all the caveats that should accompany prediction and notwithstanding the arguments made by Horton et al., it seems most likely that SGE will either be imposed on the world by a few or it will not happen at all.

SGE is a 'big' technology in that it only makes sense at scale and deployed by a state with the confidence it can maintain such large interventions well into the future. Although not especially expensive SGE does require the resources to inject sulphates, and keep injecting them, 20 km up into the stratosphere. And it does require maintaining and developing sophisticated systems of climate satellite surveillance able to monitor the effects of SGE, and adjust it accordingly. In reality, these avenues are not open to the less powerful nations of the world, and certainly not to their citizens. A 'technology of humility' (Jasanoff 2003) this is not.

Deploying and maintaining SGE implies an even greater concentration of political power than is currently the case. Whatever decision-making practices surround it, SGE ultimately requires a single metaphorical hand on the 'global thermostat'. As Clive Hamilton has quipped: SGE is "the dictator's technology of choice" (2013: 96). When it came to making choices of temperature/precipitation settings (SGE is a crude technology) it would be naïve to expect decisions to be made by the few in the interests of the many and against the interests of the few. Indeed, deployed SGE entails a concentration of power which even existing elites may baulk at: especially given that the rich are most insulated, personally, against extreme climate change and have other options including re-locating to polewards and installing climate-controlled habitations/environments.

Even the manifestly undesirable alternative of an SGE 'arms race', with different elites/powers (think Russia, China and the United States) making competing interventions into the Earth's climate implies greater centralisation of powers than currently exists (see Urpelainen 2012 for a game-theoretic analysis). In a world of rising nationalism, unilateral and competitive deployment of SGE bodes ill for the planet and democracy. As Dalby has argued, in the context of urging a rethink of geo-politics and security analysis, "the atmosphere then becomes not only a volume to be secured, but also one to be competed for directly" (2013b: 45).

In short, power in all its dimensions should be understood as *intrinsic* to SGE, rather than extrinsic. SGE presents as both a powerful technology and a technology of the powerful. 'Hubris' is the term which connects these two aspects, and indeed SGE is frequently charged with being a manifestation of hubris. SGE is a proposed technology which is more attractive to those states, classes and institutions which are currently most powerful in the contemporary world, those with the geo-political and economic might to be able to deploy it, than it is to those less powerful.

… And Also Disruptive

I am *not* arguing that existing elites will inevitably embrace SGE. There may be many practical reasons not to do so. Rather I am suggesting that SGE, both as a technology and because of the social context in which it is emerging, has a default bias towards the currently most powerful, towards the political and economic *status quo*. Clive Hamilton has called SGE a "conservative technology" (2013: 97). But this is an incomplete description. Looked at differently, SGE is a radical technology which is both conservative (in a literal sense) and also potentially disruptive of the status quo politically, economically and ontologically.

Practically, it calls into question the existing verities of climate policy, that mitigation and adaptation are the only responses to climate change. Politically, its adoption is hard to reconcile with inclusive multilateralism between states and democratic practice within states—specifically the state or states deploying SGE. It is hard to conceive of a realistic alternative democratic order, a *demos*, that could decide on what aerosol injection 'settings' to maintain, or how deployment should continue. David Victor's suggestion, as discussed in Chapter 4, is more radical and subversive than he perhaps intends: that dealing effectively with climate change means inaugurating a Brave New World of autocratic and technocratic rule. Economically, as discussed in Chapter 5, SGE appears to be hyper-compatible with capitalism, since it would appear to offer the possibility of postponing any reckoning with ecological limits (and cheaply), whilst also potentially unleashing geo-political instability. Ontologically, as we have seen in previous Chapters, SGE is especially challenging to existing assumptions about both the role and place of humans, and the human relationship to 'nature' and the more-than-human world.

Clingerman (2014) has provided a perceptive account, written from the perspective of theological anthropology, of the complex and contradictory implicit accounts of what it means to be human, held respectively by proponents and opponents of geoengineering. Both appear to conceive of the human as 'in between' the natural and the artificial. Opponents of geoengineering see humans as stewards who must maintain the separation, whilst understanding "... that human knowledge is fragmentary, our abilities are situational, and thus our responsibility is inevitably provisional and must be marked by humility". Here humans are "strangely *un/natural* beings" (Clingerman 2014: 11, italics in original) not qualified to rule the skies. This is, of course, a deeply rooted, even conservative, tradition. It is a stance which can be found in the 'Un-Natural' imaginary described in Chapter 4.

Proponents of geoengineering, Clingerman argues, interpret the 'in betweenness' of humans as a reflexive engagement with the world, rendering "... the distinction between artifice and nature indistinct in an effort to undertake a restoration of the planet" (2014: 11). Environmental failures are addressed through radical humanisation, "... not by redefining the human self in terms of being 'natural,' but by rendering the climate a domesticated participant, as it were, in such human betweenness" (2014: 12).

We see, therefore, that a Janus-like quality is apparent in SGE's re-emergence. On the one hand SGE is a response emanating from the centres of global power, rooted in standard practices and epistemologies of science, and purporting to be a solution to the climate change problem. In this sense SGE is part of a hegemonic, or at least preponderant, tale about tackling climate change, in the name of modernity, using big technology. And yet it is simultaneously a subversive story which undermines a great deal in the dominant accounts of modernity, of science and of the human-nature binary. This explains in part, why the most common sociotechnical imaginary of SGE (the 'Imperial' imaginary associated with the mainstream knowledge-brokers of SGE) has, to date, failed to gain traction as a sociotechnical imaginary. A subversive story is wrapped within a hegemonic tale, to use Ewick and Silbey's resonant phrasing (1995). The oppositional 'Un-Natural' imaginary, ironically, is less subversive. It resonates with the dominant accounts of the human-nature binary in particular, and the notion of humans as 'in between' the natural and the artificial, although it does so without the political and economic assumptions found in the Imperial imaginary.

If a dominant sociotechnical imaginary of SGE is to emerge, one which facilitates its development and deployment, then it will need to engage with these tensions. Thinking speculatively, one can anticipate SGE's proponents engaging with a number of discursive strategies if they are to generate a more compelling sociotechnical imaginary of SGE.[5] These include:

- de-naturalising 'nature' and naturalising a changed role for humanity, one which normalises human intervention at scale;
- addressing 'hubris' either by recasting it as wisdom in new circumstances, or by reconceiving the project as one of 'tinkering' and adjusting in response to feedback, and thereby casting SGE as a mundane rather than a world-making technology;
- displacing concerns about legitimacy from questions of process and structure (who speaks and what deliberative processes of decision-making might be fair) to assertions about global outcomes, emphasising claims that the deployment of SGE will be for the benefit of the poorest and most vulnerable;
- developing a normative account of why the intervention by some into the climate of others/all might be compatible with, or outweigh, long-standing rules recognising the sanctity of state sovereignty;
- making the overall vision a more positive-sounding one, in which utopian imaginings outweigh dystopian anxieties, or at least one in which the affective sense of loss (of the assumed natural order) which SGE evokes is accompanied by a sense of gain and optimism for the future.

Attaching SGE to More 'Optimistic' Imaginaries

All of these discursive moves are likely to hold together in a more compelling way if they can be attached to existing accounts or ideologies which have already gained widespread social traction, and/or if they can be presented as part of something entirely novel which makes SGE both 'obvious'

---

[5] SGE's opponents would, of course, need to challenge each of these strategies and develop a counter-narrative, which already exists in bare-bones form in the 'Un-Natural' imaginary.

and desirable. My contention is that a 'successful' sociotechnical imaginary of SGE is likely to marshal existing discourses of developmentalism/human rights as well as the paradigm of the 'Anthropocene', with its assertion that humanity has entered a new and unprecedented epoch. Indeed, signs of such a trajectory are already apparent, in nascent form, in some of the accounts of SGE, including David Keith's book *A case for climate engineering* (2013).

Development

There are indications of an incipient embrace of developmentalism by geoengineering's proponents and it is a nascent feature of the Imperial imaginary of SGE. The thrust of the argument is that the poor will be hardest hit by climate change and that SGE, by reducing the risk of catastrophic events, can be expected to help the poorest nations and people to survive and get on with their primary objective: developing. This type of approach is not unique to geoengineering. Stilgoe has drawn attention to the frequency with which new technologies, such as nano-technology or genetic modification technology, are presented as innovations that will lead to great benefits for poor countries (2015: 34).

This developmentalist turn—as I will label it—is rarely visible in the institutional assessments of SGE.[6] But it can be found in a number policy-influential texts and in some of the academic literature. The argument is frequently expressed in the language of human rights intervention and foreign aid, with SGE seen as analogous to a humanitarian intervention and an aid to those who are unable to fend for themselves or whose governments are unable to fend for them. A 2008 paper presented to a Council on Foreign Relations workshop imagines geoengineering pursued unilaterally by wealthy nations to reverse or avert sea-level rises threatening to "… impose disaster on hundreds of millions of people" (Ricke et al. 2008: 1). Geoengineering here is imagined as altruism or foreign aid. David Keith too, as described in Chapter 4, has argued that crop losses from climate change would put millions at risk (2013: 138).

---

[6] The trends I discuss below have been identified, but in different ways and from different perspectives, by others. See for example Buck (2012), and Flegal and Gupta (2018).

For him, rejecting the geoengineering 'Band-Aid' might deny "a large benefit to the poor" (2013: 137).

The expected effects of SGE on food production in the tropics and on precipitation remain highly contested. The historical evidence following major volcanic eruptions, as I illustrate in the Preface, suggests major drought and famine effects, and connections have been found between late twentieth century eruptions and Sahelian drought. Some climate modeling suggests significant declines in precipitation associated with SGE (for example work by Caldeira and Wood, cited in Suarez and Van Aalst 2017: 185). Pongratz, writing with Ken Caldeira and others, present modeling showing increased crop yields under SGE (Pongratz et al. 2012). Others claim modeling shows no region would be worse off in terms of reduced intensity of the hydrological cycle (see for example Horton 2014). Whilst expected precipitation effects are contested, so too are calculations of likely consequent effects on crop yields. There are simply too many socially-determined variables to make reliable projections about food effects.[7] Reynolds, regarded as a proponent of SGE, has expressed concern that such modelling often fails to emphasise that "aggressive mitigation … would hinder economic development, including in poor countries", and expresses the need "to emphasize that [SGE] appears to hold the potential to greatly reduce climate change risks to the environment and people, particularly to the world's poor" (2014: 184).

Some opponents of SGE have been alert to the developmentalist turn. The ETC Group have expressed their fear of SGE being normalised by being presented as "foreign aid" (2013). Their stance came in response to a modelling exercise which concluded that injecting sulphur aerosols into the stratosphere in the Northern hemisphere would lead to drought in the Sahel, but injecting them in the Southern hemisphere would lead to increased rainfall there (Haywood et al. 2013).

David Keith, as we have seen, is particularly exercised by what he sees as 'rich people' rejecting 'the geoengineering Band-Aid'. For him,

---

[7] How might one calculate the increased plant production associated with more carbon dioxide and the decreased production associated with less rain? How might human societies adapt: perhaps different crops would be grown or agricultural techniques adapted? At the most fundamental level it is notoriously unreliable to read social effects off environmental changes.

geoengineering is a necessary component of a pro-poor poverty reduction strategy. His use of the 'band-aid' metaphor is, in itself, interesting. It is in the genre of geoengineering as medical intervention, only with the treatment imagined as much more benign, even trivial, compared to the standard 'chemotherapy for the planet' metaphor. The capitalisation of 'Band-Aid' presumably references the initiatives of Bob Geldof and other music celebrities in the field of so-called humanitarian intervention and development. These initiatives have been widely critiqued, most mildly as exacerbating the very problems they presumed to address (Müller 2013; Rieff 2005).

But, to interpret Keith more generously, perhaps the major point he wishes to make is that the answer to world poverty is development, that climate change threatens that trajectory, and that SGE enables the development project to continue and may help prevent any reversal of achievements to date. In a similar vein, one of the few occasions where David Victor, in his discussions of climate policy and SGE, refers to morality is where he suggests that "rich industrialized countries have a moral obligation to adopt policies that impose the least deadweight cost on the world economy" (2011: 184). This is essentially the standard and hegemonic 'development comes through economic growth' paradigm.

A variation on the incipient developmentalist turn in geoengineering puts more emphasis on humanitarian intervention. In a paper, co-authored by two people from the Red Cross/Red Crescent Climate Centre together with a prominent and policy-influential proponent of geoengineering, Jason Blackstock, it is suggested that geoengineering technologies can be seen as a humanitarian response to climate disaster. The authors regard it as analogous to the responsibility to protect (R2P) doctrine, whereby the international community is said to have a responsibility to step in when a state, as they put it, is unable or unwilling to protect its own citizens from physical harm. They link this "duty to intervene" argument to questions of liability and compensation which might arise given that "vulnerable populations ... are also likely to be those most detrimentally impacted by any negative side effects of geoengineering experiments" (Suarez et al. 2010: 3–4).

Their paper does not exactly promote SGE.[8] But they do ask whether "the international community ha[s] a 'responsibility' to explore and develop" geoengineering, since it "*might* provide a means for avoiding some of the worst climate-induced suffering" of vulnerable populations (Suarez et al. 2010: 3, emphasis in original).[9]

The R2P doctrine endorses the notion that outside intervention is required to protect citizens from genocide, war crimes, ethnic cleansing and crimes against humanity, when their own governments do not, or are complicit. As Orford has argued, the concept can be best understood as "offering a normative grounding to the practices of international executive action that were initiated in the era of decolonization and that have been gradually expanding ever since" (Orford 2011: 10). The suggestion by Suarez, Van Aalst and Blackstock is that similar authority should exist for protecting citizens from the effects of climate change. Left unexplored is why any intervention should be into the climate system rather than the state/s most responsible for climate change. For our analytic purposes, conscripting something like an R2P principle for SGE reflects a desire to find a normative basis to legitimate intervention without consent, on humanitarian grounds.

There are clear echoes here of the more generalised securitisation of climate policy, with its association with 'executive action', which has been apparent for a number of years (Dalby 2013a; Marzec 2015). Buck notes the potential to "shift climate change from being an economic or scientific problem to being a humanitarian disaster, and geoengineering could become one critical response" (2012: 264). She contrasts the humanitarian vision with the vision which sees geoengineering as enabling capitalism to

---

[8] Not surprisingly, given their Red Cross/Crescent affiliations, they use medical metaphors to ponder how to obtain the "informed consent" of seven billion subjects "before they participate in clinical trials of a geoengineering intervention?" (Suarez et al. 2010: 3–4).

[9] Suarez and Van Aalst have since argued the need to engage with SGE to ensure "humanitarian considerations are integrated into policy decisions" (2017: 183). They note that "global power dynamics are not set up to ensure that the interests of the most vulnerable are elicited, considered, and addressed" (2017: 193). They appear to favour prior agreement, since there will be many vulnerable 'losers', on appropriate loss and damage compensation ahead of any deployment. The authors, and indeed anyone familiar with the Loss and Damage talks which accompany ongoing UNFCCC negotiations, will surely know this is a highly optimistic expectation.

continue with 'business-as-usual'. Throughout, the virtue and desirability of the development project is assumed, even though this is far from universally recognised.[10] Whilst Buck juxtaposes the humanitarian and business-as-usual visions of geoengineering, one could instead stress the connectedness of the two visions: that geoengineering framed as a humanitarian necessity and attached to 'development' is precisely what may facilitate business-as-usual, legitimise unilateral intervention and assist in normalising SGE.

Human rights discourses in their current institutionalisation, as Moyn (2010) has shown, are often mobilised as a putative component of a modern, liberal-democratic, Western international order, even if they are not universally endorsed. On this reading both 'development' and, to a lesser extent, 'human rights' have been incorporated institutionally and ideationally into the dominant international order and its practices of governance. They are universalising discourses privileging a globalising worldview.

The significance of the developmentalist turn in geoengineering lies in the work it performs in helping to re-imagine SGE as a progressive and desirable social project, as utopian even. In essence it works to reconstitute SGE as virtuous and not merely as realistic, and it provides a seemingly 'ethical' foundation for pre-emptive intervention into the climate system. In particular, the pro-poor 'development' aspect provides the virtuous reason, and the humanitarian angle provides a justification for a trans-boundary intervention (in practice by the most powerful). It attempts to reframe SGE in universal terms, as something which will need to be undertaken in the interests of the world's poorest and to protect them from climate change. The developmentalist turn is an acknowledgement that SGE is not simply a technique. It is a step towards re-imagining it as a desirable entity and part of a larger project of global ordering, of worldmaking.

---

[10] Development can be understood as a product of the dominant political-economic order in the post-colonial era (Pahuja 2011; Escobar 1995). Rist defines development as "a set of practices requiring the transformation and destruction of the natural environment and of social relations with the aim of increasing the production of commodities, goods and services" (1997: 13). There is an extensive literature challenging as simplistic the notion that the pursuit of economic growth is the key mechanism through which improvements in living conditions and well-being can be achieved (Easterly 2006). And there is extensive evidence of a symbiotic relationship between the dominant political-economic order of 'market globalism' in an unequal world, and discourses and practices of both development and institutionalised human rights, on the other (Pahuja 2011).

Anthropocene

The developmentalist turn provides both a 'halo effect' and the reassurance that geoengineering enhances the universalising project which is market globalism. However, it does little to address the ontological instabilities that invariably accompany any consideration of SGE. These relate especially to the need to reimagine the appropriate relationship between the human and the more-than-human. Existing Western modernist approaches which rely on (hu)man's dominion over nature are clearly inadequate in the face of the environmental and climate condition in which we find ourselves.

As I have discussed in earlier Chapters, narratives of Mastery no longer have the purchase they once did. Even more humble notions of environmental stewardship are rooted in an anthropocentric worldview. The charge that SGE is un-natural and hubristic is frequently found whenever publics are surveyed. It can be found, as I have shown in Chapter 4, in the Un-Natural and the Chemtrail imaginaries of SGE. Proponents of the Imperial imaginary, especially its Salvation strand, are also alert to this challenge. Its leading exponent, David Keith, is attracted to a strand of post-natural environmental thinking which sees an already humanised nature, and can therefore regard environmental interventions as a reflection of wise management, rather than being un-natural or hubristic. An 'Anthropocenic turn' is becoming apparent in climate policy generally, with the IPCC's 1.5 degree report explicitly adopting the concept (IPCC 2018).[11] The idea of the Anthropocene emerges too, sometimes explicitly, in discussions around SGE.

The Anthropocene idea, literally 'The Epoch of Humans', has clearly captured the *zeitgeist*. Today, reflections on it can be found in every discipline imaginable, across the sciences, the humanities and the social sciences. The idea of the Anthropocene is founded on the proposition that the Earth is no longer in the Holocene, still officially our current geological epoch which commenced some 12,000 years ago, and that a new geological epoch has commenced where humans (as a species) are now the major driver of changes to the physical Earth systems.

---

[11] The decision to do this is somewhat surprising given that the Anthropocene concept is not a scientifically endorsed one and given the extent to which the IPCC has been attacked previously and its legitimacy questioned when it has made even the smallest scientific error in its reports.

There is no agreement on when the epoch began although those closest to the stratigraphic debates appear to be converging on a proposal of somewhere around 1950 (Zalasiewicz et al. 2015), concurrent with the beginning of the so-called 'Great Acceleration' (Steffen et al. 2015; McNeill and Engelke 2014). Whilst the Anthropocene carries the suggestion of being a scientific concept, it currently lacks the formal imprimatur of science, at least in the sense of being formally recognised by stratigraphers as a new geological epoch. Nor is it likely to be recognised any time soon (Walker et al. 2015).

Paul Crutzen, who is associated with lifting the taboo on solar geoengineering, is also closely associated with the emergence of the term Anthropocene (2002; Crutzen and Stoermer 2000). His short 2002 article in *Nature*, 'The Geology of Mankind', concludes by touching on the implications as Crutzen saw them:

> A daunting task lies ahead for scientists and engineers to guide society towards environmentally sustainable management during the era of the Anthropocene. This will require appropriate human behaviour at all scales, and may well involve internationally accepted, large-scale geo-engineering projects, for instance to 'optimize' climate. (2002)

This founding statement is an especially managerialist and expertise-elevating interpretation of the Anthropocene. It is one still commonly encountered. For our purposes, it is important to note that for Crutzen it was apparent that geoengineering and the Anthropocene must be thought about together.

As I have argued elsewhere, the Anthropocene is typically understood to be both an *is* and an *ought*: a description of a new reality, and also a prescription as to what this implies (Baskin 2019). The concept is typically attached to one of a number of distinct normative perspectives and sensibilities. There are proponents of a 'great Anthropocene', embracing the Promethean and the technocratic, and enthusiastic about the opportunity to shape a better world. One can also find the more modest 'good Anthropocene', in which the knowledge of human culpability imposes a duty on humanity to steer the Earth in a balanced direction. And then there is a shame-inducing 'salutary Anthropocene', which observes with horror the environmental devastation which accompanies the epoch.

Most accounts of the Anthropocene imply some sort of boundary-breaking and ontological re-ordering of long-standing assumptions about the relationship between the human and the more-than-human, between 'nature' and 'culture', and between the short timescales of modernity and the long ones of the Earth, as well as the suggestion that new understandings of agency are required. Some lament the 'end of nature'. In the scientific and popular literature acknowledgement of this new ontology is often expressed crudely. According to Crutzen and Schwägerl:

> The long-held barriers between nature and culture are breaking down. It's no longer us against 'Nature'. Instead, it's we who decide what nature is and what it will be. ... [I]n this new era, nature is us. (2011: n.p.)

According to Mark Lynas in *The God Species*, "Nature no longer runs the Earth. We do" (2011: 8). One is left asking, as Szerszynski puts it, whether the Anthropocene is "the apotheosis or the end of human exceptionalism? Is this the age of the humanisation of planet or of the planetisation of the human?" (2013: 2). In most accounts of the Anthropocene it is evidently the former. The legitimating potential of this approach for SGE, with its combination of Earth-shaping and world-making, is largely self-evident.

As I have argued elsewhere (Baskin 2015), almost all versions of the Anthropocene concept, especially in the scientific and policy literature, share four attributes, and each of these can be understood in ways which contribute to a narrative that makes SGE necessary, legitimate and understood as a 'normal' part of climate management. Firstly, the concept universalises, normalises and naturalises a certain portion of humanity as the 'human' of the Anthropocene. Humanity, considered as an entity, summons up a need for global, universal responses. If our species is responsible, as an undifferentiated species, then it makes sense to gloss over or ignore the relationship between our environmental predicament and the dominant order, as it is in institutional considerations of geoengineering.

Secondly, the Anthropocene emphasises the extent to which nature's autonomy and independence have been diminished if not lost. It thereby reshapes the way the relationship between the social and the Earth is thought. It is sometimes argued that it dissolves, or recognises as dissolved, the boundary between nature and society. But, more exactly, it acknowledges that the nature-culture boundary is a constructed one. It does not so much dissolve the boundary as relocate it, whilst retaining the traditional hierarchy. It reinserts the human into nature only to re-elevate humanity above it. Seen in this way it can be understood as a colonising move (humanity has become more powerful than nature), as part of a wider claim to be a novel and universal truth. Geoengineering too can be understood as a type of boundary claim. It entails humanity moving from existence *within* climate to assertion *over* climate. The Anthropocene concept offers to make a virtue of the boundary claim, or at least deems it to have already happened: an acknowledgement that the 'facts on the ground' reveal anthropogenic impacts. In this worldview, at the risk of extending the colonial metaphor, SGE may even have something akin to a 'civilising mission', the ideology which rationalised the colonial project.

Thirdly, the Anthropocene argument typically suggests that humanity and its planet have entered a new epoch (some time around 1950). This exceptional state, where humanity is the dominant geological force, is both unprecedented and the "new normal". Exceptionality becomes a permanent attribute: an epoch. Of course the Anthropocene concept, in its scientific incarnation, is mainly about environmental or Earth systems entering a new and exceptional state. But with nature and society so entwined then exceptionality in the political, autocratic, Schmittian sense is implied. This underpinning of the Anthropocene idea is helpful for proponents of SGE. It is widely acknowledged that SGE sits uneasily with multilateral or democratic governance.

Fourthly, at least some degree of Earth systems management and technological intervention is implied, and often explicitly called for in the Anthropocene argument. Crutzen's understanding, cited above, assumes this. Lynas puts it more directly:

> The Age of Humans does not have to be an era of hardship and misery for other species; we can nurture and protect as well as dominate and conquer. But in any case, the first responsibility of a conquering army is always to govern. (2011: 11–13)

The Anthropocene idea presumes, or at the very least makes legitimate, major interventions into the workings of the earth in order to improve and save it for 'humanity'. The facticity of the Anthropocene claim (its description) is thoroughly entwined with its normative aspect (its prescription, explicit or implicit, of what to do). Certainly this normative aspect can be couched in more or less hubristic terms. But the general effect is to locate SGE as a not especially distinctive intervention.

I acknowledge, of course, that many proponents of the Anthropocene concept oppose and do not embrace SGE (although they are more likely to argue 'not now' or not until proven). For Steffen, writing with Crutzen, and for the Stockholm Resilience Institute, the Anthropocene concept implies the need for planetary 'stewardship' (Steffen et al. 2007; Rockström et al. 2009). For Biermann the concept suggests a need for 'Earth system governance' both local and global to protect Earth systems processes (Biermann 2014). The Anthropocene concept is capable of being understood in more or less Promethean ways. But it is not simply a neutral idea capable of being mobilised badly or well. Discursively, the concept does political and ideational work. It can contribute to normalising SGE. It describes and acknowledges a new ordering of things, thereby undermining pre-existing ontological assumptions on which so much resistance to SGE relies. The effect is to naturalise, scientise and inaugurate a novel human condition. It acts to globalise and universalise, fitting comfortably with mainstream understandings of climate change as a global problem. And it works to legitimate climate interventions, even exceptional ones since these are no longer regarded as 'abnormal'.

The argument here is not that the embrace of the Anthropocene implies an embrace of SGE, although it often does. My point is simply that the Anthropocene concept lends itself to being mobilised to resolve the many ontological instabilities which accompany SGE, instabilities which account for the distaste and sense of 'un-naturalness' which have made SGE difficult to 'normalise'. The Anthropocene concept, especially in its more scientific versions, and those which understand it as 'great' or 'good', provides fertile ground for the idea of SGE to flourish.

* * *

Only a re-imagined SGE is likely to provide a foundation for the technology's acceptance and deployment. This takes us into the realm of the ideational and I have critically explored above the ways in which 'development' and the 'Anthropocene' might form part of such a re-imagining. Whilst 'development' offers virtue and the claim of altruistic intervention, the 'Anthropocene' naturalises a new ordering of the world in which SGE might be understood as natural, inevitable, legitimate and unavoidable. The Anthropocene idea (or at least a particular version of it) offers the possibility for SGE to become part of a positive vision of the future. As Szerszynski has noted, new technologies are shaped by and themselves shape social practices and relations. How they are imagined is critical, particularly in the case of SGE, a technology which does not yet exist. In imagining a technology, a world and a climate are imagined too (2017).

I have suggested that the inter-relationship of three factors—how climate change unfolds, whether existing elites come to see SGE as more stabilising than disruptive, and whether SGE can be attached to a more positive and less dystopian imaginary—will shape SGE's future. Together they will shape whether and how it emerges. At the level of imaginaries I suggest that for SGE to become normalised it will need to be reborn as a sociotechnical imaginary of the Anthropocene, when its attachments to the particular science and politics of climate change are loosened, and re-attached to a positive vision of a good and modern 'Anthropocenic' world. This will not remove the risks of the technology, but it will reframe them as so many technical hurdles for human ingenuity to overcome.[12]

References

Baskin, J. (2015). Paradigm dressed as epoch: The ideology of the Anthropocene. *Environmental Values, 24*, 9–29.
Baskin, J. (2019). Global justice and the Anthropocene: Reproducing a development story. In F. Biermann & E. Lövbrand (Eds.), *Anthropocene encounters: New directions in green political thinking* (pp. 150–168). Cambridge: Cambridge University Press.

---

[12] For critics of geoengineering, this may be a cautionary tale about embracing the Anthropocene concept.

Biermann, F. (2014). *Earth system governance: World politics in the Anthropocene*. Cambridge, MA: MIT Press.

Buck, H. J. (2012). Geoengineering: Re-making climate for profit or humanitarian intervention? *Development and Change, 43*(1), 253–270.

Buck, H. J. (2018). Perspectives on solar geoengineering from Finnish Lapland: Local insights on the global imaginary of Arctic geoengineering. *Geoforum, 91*, 78–86.

Clémençon, R. (2016). The two sides of the Paris Climate Agreement: Dismal failure or historic breakthrough? *Journal of Environment & Development, 25*(1), 3–24.

Clingerman, F. (2014). Geoengineering, theology, and the meaning of being human. *Zygon, 49*(1), 6–21.

Crutzen, P. J. (2002). Geology of mankind. *Nature, 415*(6867), 23.

Crutzen, P. J., & Schwägerl, C. (2011). *Living in the Anthropocene: Toward a new Global Ethos*. Opinion piece on Yale Environment 360 website, 24 January. Retrieved January 12, 2019, from http://e360.yale.edu/feature/living_in_the_anthropocene_toward_a_new_global_ethos/2363/

Crutzen, P., & Stoermer, E. F. (2000). Have we entered the Anthropocene? Republished 31 October 2010 in *IGBP Global Change Magazine*. Retrieved January 12, 2019, from http://www.igbp.net/news/opinion/opinion/havewe enteredtheanthropocene.5.d8b4c3c12bf3be638a8000578.html

Dalby, S. (2013a). Security. In C. Death (Ed.), *Critical environmental politics* (pp. 229–237). Abingdon, Oxon: Routledge.

Dalby, S. (2013b). The geopolitics of climate change. *Political Geography, 37*, 38–47.

Easterly, W. (2006). *The white man's burden: Why the West's efforts to aid the rest have done so much ill and so little good*. New York: Penguin Press.

Escobar, A. (1995). *Encountering development: The making and unmaking of the third world*. Princeton, NJ: Princeton University Press.

ETC Group. (2013). Normalizing geoengineering as foreign aid. The Artificial Intelligence of Geoengineering Part 3. Blog post, 4 April. Retrieved January 12, 2019, from http://www.etcgroup.org/content/normalizing-geoengineering-foreign-aid

Ewick, P., & Silbey, S. (1995). Subversive stories and hegemonic tales: Toward a sociology of narrative. *Law & Society Review, 29*(2), 197–226.

Flegal, J. A., & Gupta, A. (2018). Evoking equity as a rationale for solar geoengineering research? Scrutinizing emerging expert visions of equity. *International Environmental Agreements-Politics Law and Economics, 18*(1), 45–61.

Hamilton, C. (2013). *Earthmasters: Playing god with the climate*. Crow's Nest, NSW: Allen & Unwin.
Haywood, J. M., Jones, A., Bellouin, N., & Stephenson, D. (2013). Asymmetric forcing from stratospheric aerosols impacts Sahelian rainfall. *Nature Climate Change, 3*(7), 660–665.
Horton, J. (2014). Solar geoengineering: Reassessing costs, benefits, and compensation. *Ethics, Policy & Environment, 17*(2), 175–177.
Horton, J. B., Reynolds, J. L., Buck, H. J., Callies, D., Schaefer, S., Keith, D. W., et al. (2018). Solar geoengineering and democracy. *Global Environmental Politics, 18*(3), 5–24.
IPCC (Intergovernmental Panel on Climate Change). (2001). Climate Change 2001: Synthesis report. In R. T. Watson & The Core Writing Team (Eds.), *A contribution of Working Groups I, II, and III to the Third Assessment Report of the Intergovernmental Panel on Climate Change*. Cambridge: Cambridge University Press.
IPCC (Intergovernmental Panel on Climate Change). (2013). Climate Change 2013: The physical science basis. In *Contribution of Working Group I to the Fifth Assessment Report of the Intergovernmental Panel on Climate Change*. Cambridge, UK and New York: Cambridge University Press.
IPCC (Intergovernmental Panel on Climate Change). (2018). *Global warming of 1.5°C: An IPCC Special Report on the impacts of global warming of 1.5°C above pre-industrial levels and related global greenhouse gas emission pathways, in the context of strengthening the global response to the threat of climate change, sustainable development, and efforts to eradicate poverty*. Geneva: World Meteorological Organization
Jackson, L. S., Crook, J. A., Jarvis, A., Leedal, D., Ridgwell, A., Vaughan, N., et al. (2015). Assessing the controllability of Arctic sea ice extent by sulfate aerosol geoengineering. *Geophysical Research Letters, 42*(4), 1223–1231.
Jasanoff, S. (2003). Technologies of humility: Citizen participation in governing science. *Minerva, 41*, 223–244.
Keith, D. W. (2013). *A case for climate engineering*. Cambridge, MA: MIT Press.
Lagmay, A. M. F., et al. (2015). Devastating storm surges of Typhoon Haiyan. *International Journal of Disaster Risk Reduction, 11*, 1–12.
Lynas, M. (2008). *Six degrees: Our future on a hotter planet*. Washington, DC: National Geographic.
Lynas, M. (2011). *The God species: How the planet can survive the age of humans*. London: Fourth Estate.

Marzec, R. P. (2015). *Militarizing the environment: Climate change and the security state*. Minneapolis: University of Minnesota Press.
McNeill, J. R., & Engelke, P. (2014). *The great acceleration: An environmental history of the Anthropocene since 1945*. Cambridge, MA and London: Belknap Press.
Moyn, S. (2010). *The last Utopia: Human rights in history*. Cambridge, MA: Harvard University Press.
Müller, T. R. (2013). The long shadow of band aid humanitarianism: Revisiting the dynamics between famine and celebrity. *Third World Quarterly, 34*(3), 470–484.
NASA GISS. (2016). NASA Goddard Institute for Space Studies Surface Temperature Analysis Website (GISTEMP). Retrieved January 8, 2019, from https://data.giss.nasa.gov/gistemp/
National Research Council (NRC). (2015). *Climate intervention: Reflecting sunlight to cool Earth*. Washington, DC: National Academy of Sciences.
Nixon, R. (2011). *Slow violence and the environmentalism of the poor*. Cambridge, MA: Harvard University Press.
Nyvold, M. (2015, December 23). In Greenland, hopes for climate change to boost economy. Retrieved January 15, 2019, from http://phys.org/news/2015-12-greenland-climate-boost-economy.html
Orford, A. (2011). *International authority and the responsibility to protect*. Cambridge, UK and New York: Cambridge University Press.
Pahuja, S. (2011). *Decolonising international law: Development, economic growth, and the politics of universality*. Cambridge: Cambridge University Press.
Pongratz, J., Lobell, D. B., Cao, L., & Caldeira, K. (2012). Crop yields in a geoengineered climate. *Nature Climate Change, 2*(2), 101–105.
Reynolds, J. (2014). Response to Svoboda and Irvine. *Ethics, Policy & Environment, 17*(2), 183–185.
Ricke, K., Morgan, M. G., Apt, J., Victor, D. & Steinbruner, J. (2008). *Unilateral geoengineering: Non-technical briefing notes for a workshop at the Council on Foreign Relations*, Washington, DC, May 5, 2008. Retrieved January 9, 2019, from http://www.cfr.org/content/thinktank/GeoEng_Jan2709.pdf
Rieff, D. (2005, June 24). Cruel to be kind? *The Guardian*. Retrieved January 12, 2019, from http://www.theguardian.com/world/2005/jun/24/g8.debtrelief
Rist, G. (1997). *The history of development*. London: Zed Books.

Rockström, J., et al. (2009). Planetary boundaries: Exploring the safe operating space for humanity. *Ecology and Society, 14*(2), 32.

Steffen, W., Crutzen, P., & McNeill, J. (2007). The Anthropocene: Are humans now overwhelming the great forces of nature? *Ambio, 36*(8), 614–621.

Steffen, W., et al. (2015). The trajectory of the Anthropocene: The great acceleration. *The Anthropocene Review, 2*(1), 81–98.

Stilgoe, J. (2015). *Experiment earth: Responsible innovation in geoengineering.* Abingdon: Routledge.

Suarez, P., Blackstock, J., & Van Aalst, M. (2010, March 20). Towards a people-centered framework for geoengineering governance: A humanitarian perspective. *The Geoengineering Quarterly.* Retrieved January 12, 2019, from http://www.greenpeace.to/publications/The_Geoengineering_Quarterly-First_Edition-20_March_2010.pdf

Suarez, P., & Van Aalst, M. K. (2017). Geoengineering: A humanitarian concern. *Earth's Future, 5*(2), 183–195.

Szerszynski, B. (2013). *A response to Bruno Latour's lecture "Gaia: The new body politic".* Talk delivered at 'The Holberg Prize Symposium 2013: From Economics to Ecology' conference in Bergen, 4 June. Retrieved January 12, 2019, from https://www.academia.edu/4960799/A_response_to_Bruno_Latours_lecture_Gaia_the_new_body_politic

Szerszynski, B. (2017). Coloring climates: Imagining a geoengineered world. In U. Heise, J. Christensen, & M. Niemann (Eds.), *The Routledge companion to the environmental humanities* (pp. 120–131). Abingdon, Oxon and New York: Routledge.

UNFCCC (United Nations Framework Convention on Climate Change). (2015). *Adoption of the Paris Agreement.* FCCC/CP/2015/L.9/Rev.1. Conference of the Parties, Paris, 12 December

Urpelainen, J. (2012). Geoengineering and global warming: A strategic perspective. *International Environmental Agreements: Politics, Law and Economics, 12*(4), 375–389.

Victor, D. G. (2011). *Global warming gridlock: Creating more effective strategies for protecting the planet.* Cambridge: Cambridge University Press.

Wagner, G., & Weitzman, M. L. (2015). *Climate shock.* Princeton, NJ: Princeton University Press.

Walker, M., Gibbard, P., & Lowe, J. (2015). Comment on Jan Zalasiewicz et al. 'When did the Anthropocene begin? A mid-twentieth century boundary is stratigraphically optimal. *Quaternary International, 383,* 196–203. *Quaternary International, 383,* 204–207.

World Bank. (2013). *Turn down the heat: Climate extremes, regional impacts, and the case for resilience*. A report for the World Bank by the Potsdam Institute for Climate Impact Research and Climate Analytics. Washington, DC: World Bank.

Zalasiewicz, J., et al. (2015). When did the Anthropocene begin? A mid-twentieth century boundary level is stratigraphically optimal. *Quaternary International, 383*, 196–203.

7

Conclusion

Solar geoengineering (SGE) aims to reverse or slow the rise in global temperatures by injecting aerosols into the stratosphere to reflect sunlight and reduce incoming solar radiation. It is a controversial technology. Whilst no longer 'officially' taboo, it is also not yet 'officially' accepted. It is an in-between technology. It is not yet in use but it has gone beyond being just a speculative idea. And, as discussed in the previous Chapter, it can also be understood to stand in-between the human and 'natural' worlds. SGE is a reflective technology, in two senses. By modifying the Earth's albedo it aims literally to reflect more sunlight away from the earth's surface. But it also prompts reflection on the danger of hubris and on the boundaries of what it is appropriate for humans to do. It is an imperial technology because, if adopted, it must invariably be imposed on the many by the few.

SGE is a troubling technology. It troubles because of the many climatic risks attached to it (from ozone depletion to monsoon disruption) and because of doubts that it can produce reliable, predictable climate outcomes. It troubles because it acts only to mask the warming effects of rising greenhouse gas concentrations, not tackle the concentrations themselves. It troubles because its promise to reduce or slow warming prolongs the ability for our system of extractivism and ecologically destructive growth to persist. It troubles because the very seriousness with which SGE is being discussed throws down the gauntlet to those attached to the existing verities of climate policy and mitigation-centric solutions. It troubles many of the premises of mainstream environmentalism which desperately wants to tackle climate change but is also inclined towards the (low-intervention) protection of 'nature'.

In thinking about science and technology, the ways in which we know the world are inseparable from the ways we choose to live in it (Jasanoff 2004). SGE is no exception. In thinking about SGE the material and the ideational, how the world (and climate) *is* and how it *ought* to be, are entwined. Because SGE has not (yet) been normalised or deployed, it is fruitful to examine it through the lens of imaginaries, and 'sociotechnical imaginaries' in particular (Jasanoff 2015). This helps us understand both SGE and also the work that imaginaries do in bringing new technologies into being. I have tried to explore what SGE is imagined to be, what it is imagined to be for, and for whose benefit it is imagined. All of these remain highly contested.

The interplay of relations of *power*, of *knowledge* and of *values* are critical in understanding SGE. When the 'official' institutional assessments of SGE invariably conclude that 'more research is needed', they privilege an account in which insufficient knowledge is the main impediment to SGE and more of the same knowledge is needed. And they either ignore or gloss over questions of values and power, or they implicitly embrace values which are blind to the relationships of power which infuse climate policy. In so doing they reinforce what I have labelled the Imperial imaginary of SGE.

None of the imaginaries I discuss—Imperial, Un-Natural and Chemtrail—is hegemonic. None can be regarded as 'institutionally stabilized', 'shared understandings' of SGE are not evident and all the competing imaginaries struggle to present 'desirable futures' (Jasanoff 2015: 4). These are nascent sociotechnical imaginaries. Whilst the Imperial imaginary is dominant amongst SGE's major proponents, it is not more widely embraced and indeed is strongly contested, not least in the community of climate policymakers and climate scientists. It is precisely contestation over whether and how to stabilise SGE institutionally and ideationally which is in play. For SGE's proponents this includes developing a narrative which emphasises the technology's desirable features more than its dystopian ones.[1]

In the remainder of this conclusion I will summarise the overall arguments being made and return to three questions that have recurred throughout this book. How does SGE today differ from previous incarnations of the idea? What is preventing SGE from being normalised and embraced as part of a wider strategy for tackling climate change? What is at stake in SGE's re-emergence? I will close by reflecting briefly on two questions that I am repeatedly asked when I present this work. What does the Paris climate agreement (and the election of Donald Trump as US President) mean for SGE? What does the post-natural turn in environmental thinking mean for environmentalists wanting to oppose SGE?

The Argument

In this book I have examined SGE in depth, focusing especially on the accounts provided by 'official' institutions and leading knowledge-brokers (generally proponents). Because what SGE is and what it is imagined to be has changed over time, I have adopted a simple periodisation: from 'Mastery' post-1945, through 'Taboo', to SGE's 'Re-emergence' in the mid-2000s. Chapter 2 focused on earlier efforts to promote geoengineering prior to its more recent re-emergence. In Chapters 3, 4 and 5 I took a closer look at SGE today, since the taboo on it has lifted. I unpacked the rationales and framings of contemporary SGE, the competing imaginaries which animate both proponents and opponents of SGE, and the relations of power, knowledge systems and values which have accompanied its re-emergence. Chapter 6 looked to the future, so was inevitably more speculative. It asked how SGE might unfold and how changes to climate, changing relations of power, and the embrace of discursive turns towards both the Anthropocene and developmentalism might facilitate SGE's normalisation and deployment.

---

[1] I do not mean to suggest that such narratives are developed cynically, although they may be, nor that they are not sincerely believed to be true.

I showed how particular conceptions of and assumptions about society, about climate, and about 'nature' are intimately tied up with SGE's re-emergence. SGE's re-emergence, in turn, is both challenging and reshaping some of these assumptions. Social values are shaping (and constraining) the technology, and the technology is challenging and compelling reflection on existing social values. I understand SGE as a sociotechnical project, rather than the more typical characterisation of it as a putative technology for tackling climate change which may have social and ethical implications. Social and power relations, I argue, are intrinsic to SGE. They are not additional, add-on, complications. Unusually, and distinct from nuclear technology, biotechnology and other technologies which raise comparable existential questions, SGE is being assessed prior to becoming operational, rather than alongside or following its deployment.

What's Different?

In what ways does geoengineering's re-emergence today differ from earlier iterations and imagined interventions? I describe these earlier iterations in Chapter 2. During the period of 'Mastery' no clear distinction was made between 'weather' and 'climate' and there was little or no sense of an emerging warming trajectory. Geoengineering was embraced both because it was possible, and as an assertion of power, not because of any climate or weather problem needing to be addressed. For almost three decades after the end of World War Two, both the imagining of geoengineering (and terraforming), and the very real interventions to manipulate major weather systems, formed part of Cold War efforts by the two major powers to control space, the atmosphere and the world.

From the 1970s, in both East and West, there was pushback from both scientists and from wider publics against many of the assumptions and the hubris of the Mastery era. There was growing resistance to the idea that active domination and control of nature was desirable or even possible, and increasing acknowledgement of environmental degradation and even ecological limits. This 'environmental turn' helped push consideration of geoengineering into abeyance. There was no imagined problem needing a geoengineering solution. On the few occasions the idea arises it comes to be regarded as, in effect, taboo.

Through the 1970s and 1980s one sees a growing awareness and understanding among scientists of the global warming implications of rising CO_2 emissions. By the late 1980s and early 1990s climate change started to be understood more widely, and presented as a global 'problem' requiring a global 'solution'… in particular the cutting of CO_2 emissions. There was now an acknowledged problem and also a ready solution (mitigation), although recognition of the problem's urgency was not yet widespread. This, together with the persistence of 'greener' values consistent with the 1970s environmental turn, as well as a degree of optimism in the West engendered by the end of the Cold War, helps explain why nothing came of the attempts by some in the United States to revive official support for SGE in the early 1990s and again in 2001. This failure can also be attributed to the lack of credible advocates of solar geoengineering: SGE's champions at the time were all associated ideationally with the now *passé* Mastery imaginary, and institutionally with the military-scientific-industrial complex in the US. Only in the mid-2000s, with the climate condition deteriorating and mitigation efforts manifestly underwhelming, do credible scientific proponents for lifting the taboo on SGE emerge.

In comparing geoengineering's post-War past to its post-2005 present there is clearly both an evolving understanding of climate, as well as a widely-felt attitudinal shift (not least amongst climate scientists) away from the idea that human mastery over and subjugation of nature is desirable (let alone possible). Further, since the early 1990s, a growing perception of climate as an urgent and major global problem becomes evident.

When considering geoengineering's past then, we see shifts and developments geo-politically, ideationally and in thinking about climate change specifically. But between geoengineering's past and its contemporary re-emergence there are also deep *continuities* in the assumptions made. To deploy SGE, to make it an operational technology, is to assume that the climate system *can* be 'mastered', 'managed' or 'engineered'. It is to assume that the power to shape climate in intended directions is achievable. It is to assume a human-nature relationship in which the human is both dominant and entitled to determine the shape, meaning and direction of everything else. SGE's proponents assume an imperial identity, at least implicitly, when they take on the mantle of the 'we' who must 'do something' about global climate. These assumptions hold even when they are tempered, when, for example, 'unknowns' are recognised, or the inability to determine precise outcomes is acknowledged. The more responsible and cautious proponents of SGE, when they worry about governance are, in effect, seeking a higher authority to confer upon them the managerial role.

In short, SGE's re-emergence must still rely upon many of the core assumptions and hubris of the earlier Mastery phase of geoengineering. But these assumptions are no longer regarded as self-evident, as they once were. Nor are they socially acceptable. Since the environmental turn of the 1970s they have been contested and opposed at least as much, if not more, than they are accepted. In geoengineering's earlier Mastery phase these assumptions were actively embraced by both powerful institutions and powerful scientists, and attached to optimistic visions of 'progress', self-confidence and faith in both science and scientists. But such visions are now contested, not least within the epistemic community of Earth and climate scientists. Today, whilst advocates of SGE are tied to Mastery-like assumptions regarding what *can* be done, it is no longer taken as self-evident that these actions *ought* to be done. Nor is it easy to summon up optimism in relation to the climate prognosis. The result is that SGE is hard to convey optimistically or be attached to utopian visions. As a result, contemporary proponents of SGE are driven towards a dystopian, pessimistic stance. SGE must be pursued as a 'lesser evil', a 'bad idea whose time has come', a development 'unfortunately necessary' to avert or delay 'climate catastrophe'.

What's Constraining SGE?

SGE is now part of mainstream climate policy discussion. It is respectable to talk about but it has not yet been officially embraced in IPCC reports or UNFCCC agreements. Even the major institutional assessments, discussed at length throughout this book, have been reluctant to openly embrace SGE. It has not been 'normalised', by which I mean it is not yet regarded as a valid addition to mitigation and adaptation as a climate policy option. Nor is it currently treated as a 'normal' technology needing development, as for example carbon capture and storage (CCS) is. SGE is partly accepted as an idea but not yet as a practice. Indeed, it has many vocal opponents, some arguing that its embrace is premature since there are too many unknown risks. Others argue for the turn to geoengineering itself to be resisted. But why are the arguments of SGE's proponents not gaining policy traction? What is currently restraining the normalisation of SGE as a legitimate third leg of climate policy, alongside mitigation and adaptation?

Five inter-related dynamics are working to constrain SGE's normalisation. None of these, on its own, would necessarily be fatal. But together they provide a powerful resistant dynamic.

Firstly, as I discuss at length in this book, there are obvious *uncertainties and risks* regarding the likely socio-economic, geo-political, ecosystem and physical Earth effects surrounding SGE. Much depends on the amount of warming any SGE initiative intends to mask and therefore the 'dose' of sulphate aerosols applied, as well as on the location where the injection takes place (with the southern hemisphere tropics typically suggested as ideal). 'More research' may result in better modelling and sharper, more fine-grained, projections and probabilistic estimates regarding the physical effects of SGE (regarding precipitation patterns, ozone depletion and so on).[2] But it will not produce reliable understandings of the likely socio-economic and geo-political effects. And reading environmental effects off the models is also a deeply flawed exercise. It amounts, as discussed in Chapter 5, to 'climate reductionism' the mode of analysis

---

[2] 'More research is needed' is a regularly repeated trope in the institutional reports on SGE. In practice it is hard to disentangle researching SGE from 'doing' or deploying it and calls for 'more research' can easily mask experimental deployment. There is widespread agreement that effective governance should precede research. But what if SGE is essentially ungovernable other than autocratically, as Hulme (2014) has argued—plausibly, in my view? Why research something when, if the research generates a working technology, it cannot conceivably be governed democratically? Not surprisingly, engaging with *where* (field or laboratory) and *how* to legitimately and ethically progress research into SGE, when there is insufficient agreement as to *whether* SGE is a legitimate intervention into the climate, means that the research governance questions rarely end up being about research governance. Instead they become a proxy battleground for the 'whether' question, and mask the fact that the 'whether to solar geoengineer' question remains unresolved.

in which, as Mike Hulme has put it, 'climate is first extracted from the matrix of interdependencies that shape human life within the physical world', and then elevated to the role of 'dominant predictor variable' (2011: 247). It may be possible to reduce some of the uncertainty surrounding SGE's effects, but not most of it. Much is unknown and unknowable this side of experimentation and deployment. Absent a high appetite for risk SGE is not easily normalised.

A second constraint relates to *affect and dystopia*. It has proven difficult to attach SGE to a positive vision of the future. Contrast this with the normalised climate policies of Mitigation and Adaptation: despite being defensive responses to climate change both can and have been characterised as having multiple social and economic co-benefits, and as encouraging innovative approaches and technologies. Without a positive vision of this sort, normalisation becomes more difficult. Among comparable technologies only nuclear weaponry was normalised in the context of a dystopian vision, and normalisation was as a technology able to pose a credible constraining threat to a Cold War enemy rather than as a technology that should be used. Recognition of the constraint of dystopianism may also help explain why the proponents of the Imperial imaginary of SGE seem to be searching for a more optimistic narrative. As discussed in Chapter 4, the Market strand of this imaginary emphasises SGE's relatively low cost, and thereby attaches itself to the utilitarian utopia (for some) of the efficient and unconstrained market. The Salvation strand attempts optimism, but only indirectly, since any optimism is achieved only by attaching SGE to developmentalism and altruism and the claim that it is in the interests of the world's poorest.

The *slow pace of climate change* is a third constraint. Changes to the climate are unfolding rapidly when viewed from the perspective of Earth sciences and the relative climate stability which has characterised the Holocene. But, from the perspective of individuals and their societies the changes are slow, or sporadic and highly uneven. They allow time both for some adaptation and for regarding the present as the 'new normal'. Extreme changes that might appear to justify deploying SGE can already be observed, such as numerous consecutive days on the Indian subcontinent in 2016 with temperatures well in excess of 40°C, or reports that Arctic air temperatures in November 2016 were 20°C above normal

(Vidal 2016: n.p.). But unless such changes are obviously visible, extreme and recurrent, occur in summer (a warming winter is frequently welcomed), and in a country with both the technological capability and the geo-political authority to deploy SGE, then political reality means the technology is unlikely to be deployed. Changes in climate and weather patterns in New York will count for more than those in New Delhi.[3] The pace of climate change is a constraint, with the proviso that climate circumstances could, of course, transition rapidly to becoming an enabler of SGE.[4]

What could also change rapidly, but is currently a fourth constraint, is an *absence of sufficiently powerful champions* having the power to normalise SGE or make it happen without broad consent. Enthusiasm for SGE amongst leading governments seems restricted at present to Russia. I will discuss shortly whether this changes with the election of Donald Trump. SGE as climate policy appears to lack advocates from the global South. Indeed almost all SGE's key knowledge brokers would appear to be 'male and pale' inhabitants of the global North. This is significant given the claims of some advocates from the global North that SGE should be deployed mainly to protect the world's poorest. Also absent is any evidence of interest in SGE by those who would 'benefit' immediately from its deployment; low-lying island states threatened with inundation. There are many opponents of SGE within the climate and Earth sciences. Indeed, it could be said that there is, generally speaking, a values disconnect between the dominant norms of the epistemic community of climate science and those of SGE's key proponents and knowledge-brokers. This impacts on the ability of the 'official' scientific assessments, whose panels typically include a spectrum of views, to present a consensus position favourable to taking SGE forward. It results in stalemate and this is manifest in a refusal to endorse SGE, and a deferral of decision-making by calling for 'more research'. Finally, there is currently no significant or even nascent industry with sufficient power or vested interests to champion SGE. The nexus of patents, markets and private sector players

---

[3] In an earlier Chapter I address the geo-political reasons why unilateral deployment by countries other than the USA, and possibly China, is unlikely, absent a turn to competitive SGE which would not deliver desirable climate results for anyone.
[4] Although by then it may be too late for SGE to be effective.

that does exist is too small to have significant influence on policy. Whilst a number of countries have research and development programmes, funding remains relatively limited.

Finally, SGE has a paradoxical relationship to the currently dominant order, a point I will return to shortly. On the one hand SGE appears to have the potential to cause minimal disruption to the existing economic order in the quest to address climate change: it has the capacity to mask both the warming and the driver of that warming (the endlessly expanding economy). On the other hand SGE opens up a range of unpredictable and potentially highly disruptive geo-political instabilities. In short, SGE's compatibility with the global *status quo* is not straightforward. This helps explain *both* the current lack of economically powerful champions for SGE, as well as the interest in some forms of geoengineering shown by a number of the global economic hyper-elite, such as Bill Gates or Richard Branson.

These five factors, when taken together, constrain the normalisation of SGE. The technology is currently on the cusp of rejection or acceptance. Opponents are not able to ban it or return it to the status of taboo. Proponents are not able to bring about, through persuasion, its normalisation and adoption as a third leg of climate policy.

What's at Stake?

I have explored what assumptions, interests and imagined futures travel with SGE as it re-emerges. I have unpacked the interplay of values, knowledge and relations of power that constitute it. I have shown that what is at stake is intimately wrapped up with how SGE is understood.

SGE entails far more than climate-tweaking using a mundane technology as some leading proponents imply. It is not simply, as most institutional assessments assume, a technology or *technoscientific object*, albeit with some acknowledged uncertainties and with additional social and ethical concerns attached. Rather, it is a *sociotechnical project* and the social, the ethical and the uncertain are *intrinsic* to it.

SGE is simultaneously a developing technology, an emergent climate policy, an idea-in-formation, and a claim on how the future *ought* to be. This is why understanding the competing *sociotechnical imaginaries* of

SGE, and how they might evolve—the 'Imperial' imaginary and oppositional imaginaries (such as the 'Un-Natural' and 'Chemtrail' imaginaries)—helps reveal what is at stake. Geoengineering acts as a screen onto which competing visions of the world are projected. Projected visions of the technology both shape it and enable it to (or prevent it from) coming into being. The world/s which are imagined (socially, environmentally, geo-politically, ontologically) are as important to the technology's emergence as 'nuts and bolts' considerations about the technique of SGE.

What is at stake in SGE's re-emergence, since the mid-2000s, as an object of mainstream climate policy consideration? Competing *values* are crucial here, including assumptions about who and what is important, and about what values and orderings should be retained and what not. Central, too, are the constellations of interests and relations of *power* in the world, the hierarchies and structures of the social, economic and geo-political order. Important too are the *knowledge* practices, value assumptions and power relations that shape the understanding of SGE.

Knowledge

The ways in which SGE is approached analytically in the major institutional assessments of SGE, is inadequate to the issues being addressed. Approaches which emphasise measurement and modelling (a standard practice in climate science) and elevate cost-benefit ways of seeing and valuing the world, are ill-suited to understanding and evaluating SGE. Indeed, adopting such approaches contributes to the normalisation of SGE as another leg of climate policy, and reinforces the Imperial imaginary of SGE. The knowledge hierarchies surrounding SGE elevate the physical sciences and devalue the social sciences. They also elevate expertise in general, or a particular understanding of expertise, and diminish public and experiential, grounded knowledge about climate. All this may be appropriate if SGE is 'merely' a technology or a technoscientific object. But if it is understood as a sociotechnical project, a re-making of climate and an inauguration of new world/s, then such knowledge hierarchies exclude and also conceal more than they reveal.

Calls for 'more research' before proceeding to deploy SGE presume more research in the same epistemological vein as that which has already been undertaken. In short, the ways in which expert knowledge is mobilised is not up to the task of deciding what actions to take when values are contested, in a context of deep uncertainty, and with the distinct possibility that much about SGE and climate is not only unknown but perhaps unknowable.

Values

In one sense, SGE shines a light on the unfolding of the project of Modernity. Embarking on SGE entails humans and human societies entering into a new relationship with the Earth. Whilst many in both the global North and South would agree that a new relationship with the Earth is precisely what is needed, they rarely have in mind the relationship which SGE ushers in. A great deal is at stake, therefore, because to embrace SGE is to embrace, or at least accept, an intensification of human pre-eminence in shaping the world. It is to re-embrace the practices of Mastery. It is to accept that there are no in-principle limits to technological intervention in the Earth system, that there is nothing that cannot be managed or 'fixed'. It is, in effect, to ratify rather than resist the 'End of Nature', to use McKibben's term (1989).

There are some, indeed, who welcome this direction, and perhaps it amounts to little more than the end of innocence: not for nothing is eco-modernist Mark Lynas' book titled *The God Species* (2011). But there are many who do not. Embracing, or even normalising, SGE involves choosing a particular path with the many lock-ins (multi-decadal or millennial?) required and path dependencies that are likely to ensue. It makes different, more humble, futures difficult to imagine and even harder to realise.

SGE cannot return the world to climates past. It can only manufacture a new and cooler global average climate, or a more slowly warming one. We cannot reliably predict the effects in one locale or one community or ecosystem. But we can expect climates that would be much cooler in some places than in others, and much drier or stormier or different to the present in some places than in others. Creating new weather systems at the local level—the level at which they are experienced by actual communities of people—means breaking patterns of relationship to the cycles of the natural world which shape cultures, values and material practices, and creating new ones. This is especially the case with the half of humankind still directly and extensively interacting with the land, with agriculture and with the natural world.

What other ways of being in the world, what ontologies, are erased when climate change is cast as a global problem requiring a global, universal, technological response? The worlds envisaged in the imaginaries I analyse in Chapter 4 are very different. Even so these are all imaginaries originating from and, with the partial exception of the 'Un-Natural' imaginary, largely situated in Western traditions of understanding modernity. SGE aims to make particular climates. In the process it remakes worlds, in the sense of re-shaping both the actual environments and how the world is commonly imagined to be. This is the source of much of the discomfort with SGE amongst both publics and experts.

In terms of values, a great deal is at stake: not only the implications of the path taken if SGE is embarked upon, but also the alternatives foreclosed, and the enforced restructuring of the human and more-than-human relationship in compliance with a particular perspective of Western modernity. When the 'official' assessments of SGE acknowledge that it raises ethical concerns, the value concerns with SGE are ignored. Instead 'ethics' is typically reduced to the so-called problem of 'moral hazard' (that SGE might slow down mitigation efforts) and governance. Arguably these are not ethical questions at all, but questions of power.

Power

It is argued, and is a feature of the Imperial imaginary, that without SGE the climate condition will be even worse than with SGE. This is sometimes framed in terms of 'winners' and 'losers'. Bunzl has suggested six billion winners and 1 billion losers (cited in Wood 2009) although it is unclear how this calculation is arrived at. The suggestion by SGE's proponents that without SGE things will be even worse, and the 'losers' and losses even larger in the aggregate, is hardly likely to be compelling to those negatively affected. What possible evidence would be provided for it, and by whom? And what possible real-world experience suggests such evidence would persuade the 'losers'? Recognition of this reality can be found in the widespread acknowledgment by both proponents and opponents of SGE that its deployment would likely be geo-politically destabilising. Examples of possible geo-political effects discussed in earlier Chapters include potential conflicts between nuclear-armed neighbours India and Pakistan, associated with the actual (but impossible to predict) weather effects which may materialise in one part of the Indian subcontinent rather than another. It is highly plausible, therefore, that SGE would provoke geo-political conflict. Such situations typically favour the already most powerful states.

Even leaving aside arguments that democracy and SGE are incompatible, it is hard to avoid the recognition that SGE is either ungovernable or would require autocratic governance, if only to set the desired temperature and determine the aerosol dosage. At best, 'club' governance, as proposed by David Victor, might enable concurrence by a slightly larger group of states with the geoengineering actions of the most powerful states, namely the USA and China. Within the currently most globally powerful state, the USA, it is hard to see how the desire of Florida for more cooling and the Pacific Northwest for less, could be resolved *both* prudently and democratically, not to mention responsibly in respect to other states and societies. What is at stake with SGE therefore, when taking the governance and the geo-political concerns together, is the intensification of existing concentrations of state power at best, or the emergence of new undemocratic authority, perhaps alongside the argument for 'exceptionality'. The 'Geo-management' strand of the Imperial imaginary, as discussed in Chapter 4, acknowledges this implication when it explicitly imagines a 'Brave New World'.

I have explored the paradoxical relationship between SGE and the dominant order of market globalism. As the 'Market' strand of the Imperial imaginary emphasises, the deployment of SGE is manifestly compatible with the continuation of 'business as usual'. By masking the immediate warming effects of climate change it enables the pace of mitigation to be slowed down and the costs and disruptions associated with transitioning to a low-carbon economy to be reduced or deferred.[5] In this sense, as I have argued throughout, SGE can be understood as a *doubly-masking* technology since it has the capacity also to mask the socio-economic drivers of climate change. It has the potential to mask the pursuit, endorsed by most governments and businesses and the major economic multilateral institutions, of unlimited growth in extraction, production, trade and output. SGE masks, in short, some of the central drivers of contemporary capitalism (and, historically, of Promethean communism too).

And yet, since SGE appears to require an ultra-centralised entity to deliver it, it doesn't fit neatly with environmental approaches which work with the grain of existing institutions and arrangements. Whilst SGE has the character of a technical solution to an ecological problem it seems also to imply the disruption of existing institutions. SGE would appear, therefore, to be 'hyper-compatible' with the dominant order. Its compatibility is accompanied by disruptive characteristics. This can make SGE unattractive to those elements of the dominant order for whom disruption is threatening and the predictability and continuity of existing structures and investment horizons is important. One can add here the observation that since the richest and most powerful inhabitants of the Earth have the greatest capacity (by moving, cocooning and air conditioning) to put up with even extreme climate changes, they may be the least interested in the disruptions entailed by SGE.

In short, what is at stake is the possibility of dangerous levels of geopolitical instability, the almost certain reduction in democratic accountability which would accompany the deployment of SGE, and the perpetuation of the assumption that concerns about global warming must always be secondary to the pursuit of endless economic growth.

* * *

In the most general sense what is at stake with SGE are the values we choose to embrace, the ways of knowing both the world and our ecological condition that we choose to guide us (and relatedly the role that we

---

[5] Seen from an ecological Marxist perspective, SGE enables the crisis generated by the contradiction between the Earth's limits and capitalism's need for continuous accumulation to be resolved, or at least deferred (Surprise 2018; Mann and Wainwright 2018).

assign to science and technology), whether we look at the future with hubris or humility (and thereby the policies and practices we choose to adopt), the value we place on democratic striving and inclusive, less unequal, societies, the priority we give to the pursuit of GDP growth, and our normative account of the proper relationship between the human and the more-than-human on this planet.

SGE's Shadow Presence in the Paris Agreement

The Paris climate agreement of December 2015 committed future state signatories to two main goals: holding warming to 2°C above pre-industrial, and even aiming for 1.5°C; and committing to zero net emissions in the second half of the century (UNFCCC 2015). What are the implications of the Paris agreement for geoengineering and SGE, and vice versa? The short answer is that the Paris targets make geoengineering and SGE more likely … if there is serious intent to meet the targets. There is little prospect the temperature targets can be met without immediate mitigation efforts which are dramatically stronger than existing commitments … other than by deploying SGE. And the goal of reaching zero emissions relies on what the agreement repeatedly refers to as 'anthropogenic emission removals', in effect by utilising some form of geoengineering of the carbon-dioxide removal (CDR) variety, sometimes known as 'negative emission technologies'.

SGE is not mentioned in the Paris agreement. And no specific CDR technologies are endorsed. But only one of the IPCC pathways (RCP 2.6) comes close to being aligned to the goals contained in the Paris agreement (Beck and Mahony 2017) and this assumes that meeting a zero emissions target requires the massive expansion and uptake of Biomass Energy with Carbon Capture and Storage (BECCS). This is a technology that involves growing very large quantities of biomass, burning it to produce energy and sequestering the associated CO_2 underground or in the deep oceans. BECCS comes with many challenges: it is still largely an experimental technology, it is expensive and energy intensive and comes with scale-up challenges such as the building of an extensive transport and storage infrastructure. Importantly, BECCS comes with substantial social, land-use and justice implications (Minx et al. 2018; Anderson and Peters 2016). It would require plantations and infrastructure to be developed (most effectively in the tropics of the global South) across a land area almost the size of Australia; or put another way, an estimated one-third of the world's arable land would need planting for BECCS use by 2100 (Williamson 2016). In short, the ambitious and admirable targets set in Paris can equally and plausibly be seen as faith-based policymaking. Kruger et al. (2016) have called it an example of "magical thinking", the belief that "thinking about something amounts to doing it, even when the instrumental means to bring it about are absent" (Rayner 2016: 1). It can also be understood as a form of colonial thinking. After all, whose land is it imagined to be that would be re-assigned to producing biomass for sequestration?

It is, of course, true that the zero emissions target is some decades away and perhaps some revolutionary new negative emission technology may emerge which avoids the many problems of BECCS. But the temperature target, the so-called 'guardrail' of 2°C (and 1.5°C), is closer to hand and is arguably the primary metric of the Paris agreement. SGE offers the possibility of meeting the temperature target and, in the process, buying time to develop suitable negative emissions technologies, and to do this cheaply enough to allow resources to be channelled towards that goal. This point has not been lost on SGE's proponents: "the fact that the Agreement contains explicit quantitative temperature goals instead of, say, an explicit carbon budget goal … actually facilitates the eventual inclusion of SRM in the post-Paris system" (Horton et al. 2016).

Perversely, rising CO_2 concentrations are good for biomass growth, and since SGE masks rather than tackles these concentrations, it may even be argued that it has the co-benefit of facilitating BECCS. Whilst the Paris agreement normalises carbon dioxide removal-style geoengineering, the temperature targets make SGE's normalisation more likely—not imminently, but perhaps dependent on the perceived viability of various negative emissions technologies. SGE is absent from the Paris agreement. But it is an absent presence, an unacknowledged shadow presence within the agreement. If BECCS can be understood as expressing a form of colonial thinking because it implies the re-occupation of tropical lands, then SGE can be understood as imperial thinking because it imagines dealing with the climate problem globally and from on high.

The election of Donald Trump as US President makes the future of SGE less predictable. His stated policies, and the withdrawal of the US from the Paris agreement, make achieving the necessary global emissions reductions less likely, and therefore increase the likelihood that the 2°C threshold will be crossed. On the one hand his administration shows little interest, indeed open hostility, towards mainstream climate policy. On the other hand a number of Trump's key policy advisors both share his climate denialism/ inactivism *and* are also SGE enthusiasts. This is true of both David Schnare and Newt Gingrich, both of whose views are discussed elsewhere in this book, and of his first nominee to head up the transition team of the Environmental Protection Agency, Myron Ebell (see Ebell 2009, 2016). It is also evident when the Trump-supporting chair of the Energy subcommittee of the US House of Congress, Randy Weber, argues that "the future is bright for geoengineering" (Ellison 2018). All are hostile to climate change 'alarmism', sceptical of the need for mitigation and champions of increased fossil fuel use. They are interested in SGE as a 'just in case' solution, if the warnings about warming turn out to be true, and one which doesn't constrain capitalism. In practice such views may also align with those of some of the liberal technophilic elite in the US, even as they may differ over their stance towards climate change. This raises the prospect of SGE being a shared project of the US elite, Democrat and Republican alike.

In short, the Paris agreement opens a trajectory in which the deployment of SGE becomes more likely. But the hostility of the UNFCCC establishment, and most countries, towards SGE remains deep-seated, and makes SGE's official embrace unlikely anytime soon. At the same time the Trump regime means that the chances of stumbling into SGE's deployment appear to be growing.

Troubling Environmentalism

It is SGE's 'un-naturalness' and hubris that most disturbs people when they are surveyed about SGE, when advocacy groups opposing SGE make their case, and even when many scientists consider the issue. Boucher et al. (2014) are not alone in acknowledging that the idea that geoengineering is 'unnatural', even that it should be regarded as unimaginable or taboo, arises repeatedly in the growing number of studies of public perceptions and responses to geoengineering. My own experience concurs. Over the past five years I have observed a pattern in the responses I receive when explaining SGE to non-specialists in the most professional and neutral ways I can muster. I am sometimes asked 'could it work?', never asked 'how could SGE be made to work?', but invariably asked, often with an accompanying grimace, 'who imagines doing such a thing?'. This is less a reflection of the respondents' 'green-ness' or embrace of the natural world, than an inchoate sense that geoengineering somehow breaches the 'natural' order, the sense of how things should be between humans and 'nature', and what is permissible. Clearly, distaste for, and resistance to the idea of SGE itself is widespread. The oppositional Chemtrail and Un-Natural imaginaries both reflect and build on this, as well as on suspicion of the elite nature of the SGE project, albeit in different ways.

Public attachment to 'nature' and rejection of 'un-natural' climate interventions sits comfortably with long-standing conservationist and preservationist traditions in environmentalism, especially in the West. On the other hand, a notion of the 'natural' is today often dismissed by contemporary environmentalists, eco-modernists, and environmental philosophers as a value-laden and un-scientific concept, and an inadequate foundation for environmentalism (Marris 2011; Vogel 2015). The idea of nature can also sit uneasily with the scientific discourses that have been embraced by contemporary environmentalism (unlike in the 1970s), and which constitute the dominant register and discipline in which climate change is today discussed.

As a result, the more visceral anxieties about SGE being 'un-natural' and hubristic have little place to go. Arguably, they emerge indirectly, even amongst proponents, in claims that SGE is needed *for* the environment or in anxiety about 'moral hazard' making emissions reduction less likely. It is evident when the major institutional assessments of SGE assert the importance of values and ethics. It is manifest when David Keith, in his book *A case for climate engineering* (2013), takes great pains to stress his personal love of the outdoors and connection to nature. It can be seen in the more sceptically-inclined climate scientist Alan Robock's listing of

26 risks and five benefits of SGE (2015). These include 'beautiful red and yellow sunsets' as one of the benefits, and among the risks includes, in addition to expected ones such as ozone depletion and ocean acidification and irreversibility, that there would be 'whiter skies', 'no more Milky Way' and 'do we have the right to do this?' (2015: n.p.).

There is no doubt that 'nature' is a slippery concept, perhaps the most complex in the English language, as Raymond Williams has argued, containing 'often unnoticed, an extraordinary amount of human history' (1983: 64). And as I have argued elsewhere (Baskin 2015), many issues are enmeshed in the term, from the empirical 'nature' in the world 'out there', to the concept of 'Nature' with its history, politics and complexity, to the associated Nature-Culture binary which frames much of Western thinking and scientific practice (cf. Merchant 1980; Cronon 1995; Latour 2004). It is today widely acknowledged that humans and their societies exist in, not above or alongside, 'nature', and that the popular understanding of what is, and is not, natural, shifts over time and is valued variously.

This suggests that 'nature' and a sense of SGE being 'un-natural' may be a weak conceptual foundation on which to rest opposition to SGE. More broadly SGE may provide pause for environmentalists and environmentalism more generally, to re-think a number of their existing standpoints. It is hard to proclaim the 'End of Nature' (McKibben 1989) and then to oppose interventions that engage with that reality in post-natural ways. This is the standpoint adopted by Lynas (2011) when he argues that 'we' are already engineering the Earth when we drive our cars, the implication being that embracing geoengineering is nothing really new. The response to Lynas, that geoengineering is different because it is undertaken intentionally, is manifestly a weak one: each new decision to drive or to build a car plant is intentional and taken in a context where the ecological effects are now known.

A great deal of thought is going into re-thinking environmentalism. Some aim to acknowledge the human shaping of 'nature' and urge that we treat the wilds as a 'rambunctious garden' and recognise it is no longer an untamed 'other' (Marris 2011). Some suggest the need for a new environmental theory of the post-natural world, rooted in an acknowledgement that nature has ended (Vogel 2015) even as environmental problems persist. Others, including some of SGE's advocates and sympathisers, have hoisted the banner of the Breakthrough Institute's version of ecomodernism (see Ecomodernist Manifesto 2015) which not only acknowledges nature's end but calls for its active modernisation. These are challenging perspectives for those wishing to oppose the SGE turn currently underway.

Against such approaches the argument can be made that attachment to 'nature' should not be lightly abandoned, not least because it attaches to popular discourse. It can be argued that artifice and wildness are only two extremes of a long and complex continuum, and that acknowledging that nature is damaged does not mean it has ended, nor that it has it lost its exteriority, agency and otherness (Worthy 2013). Perhaps nature has not ended, it was 'always already social', but rather humans have ended, in the sense that they have become more alert to the ways in which they are entwined with nature. Perhaps Western humans are rediscovering truths that other cultures and times have long appreciated, including the opportunity to re-imagine their relationship with the more-than-human world rather than simply continue to re-engineer that world to fit the human. My point here is simply that the idea of SGE, and its potential deployment and effects, can help illuminate the re-thinking of environmentalism that is currently underway, and that there are implications in embracing post-nature perspectives too rapidly.

* * *

One of my concerns throughout has been to understand what is constraining SGE from being normalised as a component of climate policy. One could flip this question and ask what is it in the practice of climate policy and climate science that gives rise to ideas such as SGE? Is it possible that the currently dominant approach to climate policy, combined with the climate prognosis, is now able to generate only dystopian imperial ideas like SGE or magical, colonial, thinking like BECCS? The limitations of understanding climate change mainly through the lens of science become apparent when looking at SGE. Reflecting deeply on SGE invites us to rethink how climate change is understood and approached more generally.

Certainly, it can be argued that from the 1980s understanding climate became increasingly detached from the social conditions of existence. Average weather (climate) was privileged over lived weather and experienced seasons. Heavy reliance on surveillance from above for our data, the privileging of calculative expertise, the casting of climate as a 'global' problem, all have arguably contributed to this detachment. Further, as Fleming and Jankovic have noted, when the 'social' is then re-attached it is typically in the form of something determined by climate change: "[it] 'impacts' the economy, 'affects' countries, 'harms' national security, 'hurts' the world's poor, and potentially 'leads' to global conflict" (2011: 3). Another effect has been, as Crist points out, that the attention given to climate change has had the effect of displacing other ecological concerns and de-emphasising, perhaps unintentionally, their connectedness (2007).

In parallel, a sense of the numinous when thinking about climate and weather (and indeed about 'nature') has been eroded. And yet the numinous, the sense of being in the presence of something awe-inspiring and 'other', not necessarily in a religious sense, has long been a critical component in common understandings of climate (and 'nature'). Would a science of climate that was both more down to Earth and more connected to the numinous, produce an idea like SGE? Such questions and their implicit critique are not novel. But reflecting on SGE perhaps brings them to the fore anew.

SGE is a technology which is both *climate-shaping* and *world-making*. For SGE to happen it will need to be normalised either by being presented and accepted as optimistic and discursively coherent, or by being coercively imposed. Or both. Absent a radical turn in thinking about climate and the environment, my expectation is that SGE will be deployed coercively, in a few decades time, in response to calls to 'do something' about unruly weather. Its deployment will likely be 'too much, too late': by which I mean that the climate condition will be significantly more disrupted than currently and require a higher dosage of sulphate aerosols than is prudent. This, in turn, is likely to have effects we simply cannot anticipate and usher in both a new climate (which may take centuries to stabilise around a new equilibrium), and with it, new World/s.

I acknowledge that this is a dystopian vision. But it is possible to combine pessimism of the intellect with optimism of the will, and even to have hope without optimism. And my hope is that better alternatives than SGE may emerge, that resistance to its deployment is possible, and that more humble ways of being in the world are possible too.

References

Anderson, K., & Peters, G. (2016). The trouble with negative emissions. *Science, 354*(6309), 182–183.
Baskin, J. (2015). Paradigm dressed as epoch: The ideology of the Anthropocene. *Environmental Values, 24*, 9–29.
Beck, S., & Mahony, M. (2017). The IPCC and the politics of anticipation. *Nature Climate Change, 7*, 311–313.
Boucher, O., et al. (2014). Rethinking climate engineering categorization in the context of climate change mitigation and adaptation. *WIREs Climate Change, 5*(1), 23–35.
Crist, E. (2007). Beyond the climate crisis: A critique of climate change discourse. *Telos, 141*(Winter), 29–55.
Cronon, W. (1995). *Uncommon ground: Rethinking the human place in nature.* New York: W.W. Norton & Co.
Ebell, M. (2009). The anti-green ecologist. Retrieved January 9, 2019, from https://cei.org/op-eds-and-articles/anti-green-ecologist
Ebell, M. (2016). Myron Ebell: Promoting sound policy for real environmental improvement. Retrieved January 15, 2019, from https://cei.org/blog/myron-ebell-promoting-sound-policy-real-environmental-improvement
Ecomodernist Manifesto. (2015). Retrieved January 9, 2019, from http://www.ecomodernism.org/
Ellison, K. (2018, March 28). Why climate change skeptics are backing geoengineering. *Wired.* Retrieved January 9, 2019, from https://www.wired.com/story/why-climate-change-skeptics-are-backing-geoengineering/
Fleming, J. R., & Jankovic, V. (2011). Revisiting Klima. *Osiris, 26*, 1–16.
Horton, J. B., Keith, D. W., & Honegger, M. (2016). Implications of the Paris agreement for carbon dioxide removal and solar geoengineering. Harvard Project on Climate Agreements viewpoint paper, July. Retrieved January 9, 2019, from https://www.belfercenter.org/sites/default/files/files/publication/160700_horton-keith-honegger_vp2.pdf

Hulme, M. (2011). Reducing the future to climate: A story of climate determinism and reductionism. *Osiris, 26*, 245–266.
Hulme, M. (2014). *Can science fix climate change? A case against climate engineering.* Cambridge: Polity Press.
Jasanoff, S. (2004). The idiom of co-production. In S. Jasanoff (Ed.), *States of knowledge: The co-production of science and social order.* London and New York: Routledge.
Jasanoff, S. (2015). Future imperfect: Science, technology and the imaginations of modernity. In S. Jasanoff & S.-H. Kim (Eds.), *Dreamscapes of modernity: Sociotechnical imaginaries and the fabrication of power* (pp. 1–33). Chicago: Chicago University Press.
Keith, D. W. (2013). *A case for climate engineering.* Cambridge, MA: MIT Press.
Kruger, T., Geden, O., & Rayner, S. (2016, April 26). Abandon hype in climate models. *The Guardian.* Retrieved January 15, 2019, from https://www.theguardian.com/science/political-science/2016/apr/26/abandon-hype-in-climate-models
Latour, B. (2004). *Politics of nature: How to bring the sciences into democracy.* Cambridge, MA: Harvard University Press.
Lynas, M. (2011). *The God Species: How the planet can survive the age of humans.* London: Fourth Estate.
Mann, G., & Wainwright, J. (2018). *Climate Leviathan: A political theory of our planetary future.* London: Verso.
Marris, E. (2011). *Rambunctious Garden: Saving nature in a post-wild world.* New York: Bloomsbury.
McKibben, B. (1989). *The end of nature.* New York: Random House.
Merchant, C. (1980). *The death of nature.* New York: HarperCollins.
Minx, J. C., Lamb, W. F., Callaghan, M. W., Fuss, S., Hilaire, J., Creutzig, F., et al. (2018). Negative emissions—Part 1: Research landscape and synthesis. *Environmental Research Letters, 13*(6), 63001.
Rayner, S. (2016). What might Evans-Pritchard have made of two degrees? *Anthropology Today, 32*(4), 1–2.
Robock, A. (2015, September 1). A modest proposal. *Huffington Post.* Retrieved January 15, 2019, from https://www.huffingtonpost.com/alan-robock/a-modest-proposal_15_b_8059256.html
Surprise, K. (2018). Preempting the second contradiction: Solar geoengineering as spatiotemporal fix. *Annals of the American Association of Geographers, 108*(5), 1228–1244.

UNFCCC (United Nations Framework Convention on Climate Change). (2015). *Adoption of the Paris Agreement.* FCCC/CP/2015/L.9/Rev.1. Conference of the Parties, Paris, 12 December.
Vidal, J. (2016, November 22). 'Extraordinarily hot' Arctic temperatures alarm scientists. *The Guardian.* Retrieved January 15, 2019, from https://www.theguardian.com/environment/2016/nov/22/extraordinarily-hot-arctic-temperatures-alarm-scientists
Vogel, S. (2015). *Thinking like a mall: Environmental philosophy after the end of nature.* Cambridge, MA and London: MIT Press.
Williams, R. (1983). *Keywords: A vocabulary of culture and society* (Rev. ed.). London: Fontana.
Williamson, P. (2016). Scrutinize CO_2 removal methods. *Nature, 530,* 153–155.
Wood, G. (2009, July/August). Re-engineering the Earth. *The Atlantic.* Retrieved January 15, 2019, from http://www.theatlantic.com/magazine/archive/2009/07/re-engineering-the-earth/307552/
Worthy, K. (2013). *Invisible nature.* Amherst, NY: Prometheus Books.

Index[1]

A

American Enterprise Institute (AEI), 99, 130, 132, 165n1, 175, 176, 177n9, 188, 189, 191
Anthropocene, 4, 21, 86, 111, 214, 224, 229–234, 229n11, 243
Arctic Methane Emergency Group (AMEG), 96, 97, 97n12, 182

B

Bickel, Eric, 99, 130, 131, 131n2, 165n1, 175, 176, 188, 191
Bipartisan Policy Center (BPC), 2n1, 6, 17, 90, 99, 103, 105n14, 133, 137n5, 165, 186, 198, 199
Budyko, Mikhail, 36, 39, 43, 49, 50, 53, 60, 93n11

Bush, George W, 55, 77, 77n2, 79, 84, 84n5, 153

C

Caldeira, Ken, 18n11, 103, 196, 225
Chemtrails, 3, 8, 14, 21, 123, 124, 140–145, 149–151, 154, 155, 229, 242, 251, 259
Climate change
  Kyoto, 1, 53, 77, 133, 191
  Paris accord, 1, 6n7, 91, 182, 190, 215, 243, 256–258
  UNFCCC, 53, 57, 59, 91, 182, 215, 256
Climate engineering, *see* Geoengineering

---

[1] Note: Page numbers followed by 'n' refer to notes.

Cold War, 11, 20, 27–30, 33n2, 35, 36, 47, 48, 50, 51, 55, 56, 58–61, 62n10, 63–65, 67, 244, 245, 248
Copenhagen Consensus, 99, 165n1, 176, 177n9
Council on Foreign Relations (CFR), 6, 17, 90, 91, 186, 224

Crutzen, Paul, 5, 86–87, 98, 103, 104, 165, 230–232

E
Earth Summit, 53, 57, 59, 64, 77n2
Environmentalism, 4, 13, 15, 22, 45, 47, 48, 57n7, 58–60, 63, 65, 104, 127, 135, 139, 140, 150, 151, 154–156, 171, 174, 192, 195, 196, 229, 242, 243, 258–263
Environmental Modification Convention (ENMOD), 45, 46, 125
ETC Group, 2n1, 7, 18, 125–128, 165n1, 225
European Transdisciplinary Assessment of Climate Engineering (EuTRACE), 18, 92

F
Fedorov, Y.K., 42
Frosch, Robert, 51, 52, 56

G
Geoengineering
 and capitalism, 227
 carbon-dioxide removal (CDR), 6n7, 7, 90n9, 91, 168, 256
 and climate modelling, 9–10
 compared to other technologies, 254
 cost of, 10, 99, 144, 168, 175, 179
 and developmentalism, 224
 and emergency, 19, 76, 80, 83, 85, 96, 99–101, 103, 134, 180, 182, 183, 196
 ethics/values, 11, 17, 19, 164, 179, 196–198
 frames and metaphors, 100–104
 governance, 19, 92, 166, 177n8, 197, 254
 history of, 14, 19, 27, 29
 hubris, viii, 12, 19, 40, 127, 128, 246
 knowledge about, 60, 89
 norms/normalisation, 247
 rationales for, 20, 96, 98, 99, 101, 110, 111
 representations of, 9, 105, 106
 and risk, 98, 100, 101, 103, 109, 110, 137, 138, 171, 180
 and security, 16n10, 17, 187
 Solar Radiation Management (SRM), 91, 99, 168
 techno-fix, 127, 129
 uncertainty, 85, 164, 171
 US politics, 58, 59
 utopia/dystopia, 8

Index 269

Geoengineering Model Intercomparison Project (GeoMIP), 166
Globalisation, 16, 20, 55, 58–60, 64, 75, 76, 78n3, 129, 142, 145, 214, 229, 254
Global South, 47, 51, 63, 64, 75, 125, 128, 135, 185, 189, 219n4, 249, 257
Global warming, see Climate change
Government Accountability Office (GAO), 6, 17, 90, 98, 102, 110, 133, 137n5, 165, 181, 199

H

Hands Off Mother Earth (HOME), 18, 125, 127, 128
Hulme, Mike, 18, 101, 110, 127n1, 170, 184, 189, 247n2, 248
Huxley, Julian, 30, 30n1, 31

I

Imaginaries, see Sociotechnical imaginary
Integrated Assessment of Geoengineering Proposals (IAGP), 92
Intergovernmental Panel on Climate Change (IPCC), 1–2n1, 6, 7, 9, 17, 43, 53–55, 61n9, 79, 89–91, 91n10, 93, 93n11, 97–101, 105, 106, 111, 133, 137, 137n5, 149, 153, 165, 166, 173, 175–177, 181, 183, 195, 198, 214, 215, 229, 229n11, 246, 256
Izrael, Yuri, 53, 93, 93n11

J

Jasanoff, Sheila, 3, 8, 10, 11, 17, 124, 152–154, 163, 220, 242

K

Keith, David, 18, 18n11, 40, 53, 55, 94, 98, 103, 111, 136–140, 137n5, 150, 152n10, 153–155, 173, 183, 192, 193, 196, 224–226, 229, 259
Knowledge-brokers, 15–19, 18n11, 94, 107, 108, 152n10, 153, 163, 180, 186, 188, 189, 191, 192, 222, 224, 243, 249

L

Latour, Bruno, 150, 260
Lawrence Livermore National Laboratory (LLNL), 51, 53, 54, 57, 62n10, 105
Lomborg, Bjorn, 85

M

MacCracken, Michael, 18n11, 53–55, 57
Massachusetts Institute of Technology (MIT), 40, 48, 61, 82, 84, 99

N

National Academy of Sciences (NAS), 43, 51, 52, 56, 84, 84n5, 87
National Aeronautics and Space Administration (NASA), 6, 56, 57, 64n11, 87–90, 98, 104, 130, 176, 216
National Research Council (NRC), 2n1, 5–7, 9, 14, 16, 17, 43, 81, 88n8, 90, 90n9, 94, 95, 98, 103, 105, 106, 109, 133, 137n5, 142, 153, 155, 165–171, 174, 175, 181, 185–188, 191, 198, 200, 213n1
NOVIM, 174, 180, 199

R

RAND Corporation, 6, 17, 41, 90, 175
Robock, Alan, 5, 43, 107n15, 109, 126, 259
Royal Society, 5, 7, 16, 17, 92, 95, 96, 99, 101–103, 110, 137n5, 153, 165, 168, 169, 171, 173, 174, 174n6, 180, 188, 197, 198, 200, 201

S

Schneider, Stephen, 5, 49, 52, 53, 57, 103, 111
Schwartz, Peter, 81, 82
Sociotechnical imaginary, 4, 8–9, 123–124, 151–154, 242–243
  Chemtrail imaginary, 3, 8, 21, 123, 145, 149, 193, 242, 259
  Geo-management narrative, 21, 123, 129, 133–136, 138n7, 145, 149, 151, 152, 156, 156n11, 218, 254
  Imperial imaginary, 3, 8, 14, 21, 123, 124, 126, 129, 132, 136, 144, 145, 149–155, 163, 168, 182, 183, 192, 202, 218, 219, 222, 224, 229, 242, 248, 251, 253, 254
  Market narrative, 21, 123, 129–133, 137, 145, 149, 151, 152, 152n9, 152n10, 156n11, 218, 248, 254
  Salvation narrative, 123, 129, 136–140, 145, 149–152, 152n9, 154, 156n11, 182, 218, 229, 248
  Un-Natural imaginary, 3, 4, 8, 13, 14, 21, 123–128, 144, 145, 149–151, 154, 222, 223n5, 242, 253, 259
Stratospheric Controlled Perturbation Experiment (SCoPEx), 90, 107, 137
Stratospheric Particle Injection for Climate Engineering (SPICE) project, 2n4, 19, 92, 103, 107n15

T

Tambora, 13, 166n2
Teller, Edward, 33, 33–34n2, 34, 51, 54, 57, 64, 84–86, 84n5, 85n6, 103
Tyndall Centre, 82–86, 101

U

Umweltbundesamt, 17, 92, 105n14, 199
US Congress, 37, 132

V

Victor, David, 18n11, 19, 94, 133–137, 138n7, 152n10, 155, 188, 217, 221, 226, 254

W

Wood, Lowell, 13, 54, 55, 57, 83–85, 84n5, 84n6, 99, 166n2, 225, 253

Z

Zworykin, Vladimir, 34–36

CPSIA information can be obtained
at www.ICGtesting.com
Printed in the USA
LVHW071732290519
619458LV00008B/14/P